Germain Adriaan (Ed.)

Nokia 9500 Communicator

Germain Adriaan (Ed.)

Nokia 9500 Communicator

Smartphone, Series 90 (software platform), OMAP

Brev Publishing

Publisher:
Brev Publishing is a trademark of
International Book Market Service Ltd., 17 Rue Meldrum, Beau Bassin, 1713-01 Mauritius
Email: info@bookmarketservice.com
Website: www.bookmarketservice.com

Published in 2012

Printed in: U.S.A., U.K., Germany. This book was not produced in Mauritius.

ISBN: 978-620-0-36427-2

Contents

Articles

References

Nokia_9500_Communicator

Nokia 9500 Communicator is a smartphone produced by Nokia, introduced in 2004. It runs on the Symbian-based Series 90 platform, albeit disguised to look like the Series 80 platform.

Nokia 9500

The 9500 is equipped with a 150 MHz Texas Instruments OMAP processor and 64 megabytes of SDRAM. It features two screens – a fully functional interior screen, and a restricted-function external screen, which operates on a stripped-down Series 40 user interface.

Connectivity features of the 9500 include: Bluetooth, infrared, USB, Wi-Fi, CSD, HSCSD, GPRS and EDGE. It has send and receive fax facilities (without scanner) and multi-account POP3/IMAP email client.

The built-in Web browser, a Nokia-branded version of Opera, is able to render both WAP and HTML Web pages. Like others in the Communicator series, the 9500 has a full QWERTY keyboard.

Built-in software includes a word processor, spreadsheet and presentation program, which are compatible with the Microsoft Office suite equivalents; also featured is an MP3 player. In addition to the software applications provided by Nokia, a large range of third-party software is available; many programs written for older Nokia Communicators are compatible with the 9500 and new software can be written in C++ or OPL.

The 9500 also runs Java ME applications, but some do not make full use of the unusually large and wide screen, so that many existing Java games will run, but only use the top left hand corner of the screen. It supports storage on a Multimedia Card (MMC) of up to 2 GB.

See also

- Nokia 9210, predecessor (2002)
- Nokia 9300, a device with similar specs, except smaller and without wifi and camera.
- Nokia E90 Communicator, successor to the 9500
- List of Nokia products

Smartphone

A **smartphone** is a high-end mobile phone built on a mobile computing platform, with more advanced computing ability and connectivity than a contemporary feature phone.[1] [2] [3] The first smartphones were devices that mainly combined the functions of a personal digital assistant (PDA) and a mobile phone or camera phone. Today's models also serve to combine the functions of portable media players, low-end compact digital cameras, pocket video cameras, and GPS navigation units. Modern smartphones typically also include high-resolution touchscreens, web browsers that can access and properly display standard web pages rather than just mobile-optimized sites, and high-speed data access via Wi-Fi and mobile broadband.

Modern smartphones.

The most common mobile operating systems (OS) used by modern smartphones include Google's Android, Apple's iOS, Microsoft's Windows Phone, Nokia's Symbian, RIM's BlackBerry OS, and embedded Linux distributions such as Maemo and MeeGo. Such operating systems can be installed on many different phone models, and typically each device can receive multiple OS software updates over its lifetime.

The distinction between smartphones and feature phones can be vague and there is no official definition for what constitutes the difference between them. One of the most significant differences is that the advanced application programming interfaces (APIs) on smartphones for running third-party applications[4] can allow those applications to have better integration with the phone's OS and hardware than is typical with feature phones. In comparison, feature phones more commonly run on proprietary firmware, with third-party software support through platforms such as Java ME or BREW.[1] An additional complication in distinguishing between smartphones and feature phones is that over time the capabilities of new models of feature phones can increase to exceed those of phones that had been promoted as smartphones in the past.

History

Early years

IBM Simon (introduced 1992) shown in the charging station

The first smartphone was the IBM Simon; it was designed in 1992 and shown as a concept product[5] that year at COMDEX, the computer industry trade show held in Las Vegas, Nevada. It was released to the public in 1993 and sold by BellSouth. Besides being a mobile phone, it also contained a calendar, address book, world clock, calculator, note pad, e-mail client, the ability to send and receive faxes, and games. It had no physical buttons, instead customers used a touchscreen to select telephone numbers with a finger or create faxes and memos with an optional stylus. Text was entered with a unique on-screen "predictive" keyboard. By today's standards, the Simon would be a fairly low-end product, lacking a camera and the ability to download third-party applications. However, its feature set at the time was highly advanced.

The Nokia Communicator line was the first of Nokia's smartphones starting with the Nokia 9000, released in 1996. This distinctive palmtop computer style smartphone was the result of a collaborative effort of an early successful and costly personal digital assistant (PDA) by Hewlett-Packard combined with Nokia's best-selling phone around that time, and early prototype models had the two devices fixed via a hinge. The Communicators are characterized by a clamshell design, with a feature phone display, keyboard and user interface on top of the phone, and a physical QWERTY keyboard, high-resolution display of at least 640×200 pixels and PDA user interface under the flip-top. The software was based on the GEOS V3.0 operating system, featuring email communication and text-based web browsing. In 1998, it was followed by Nokia 9110, and in 2000 by Nokia 9110i, with improved web browsing capability.

In 1997 the term 'smartphone' was used for the first time when Ericsson unveiled the concept phone GS88,[6] [7] the first device labelled as 'smartphone'.[8]

Symbian

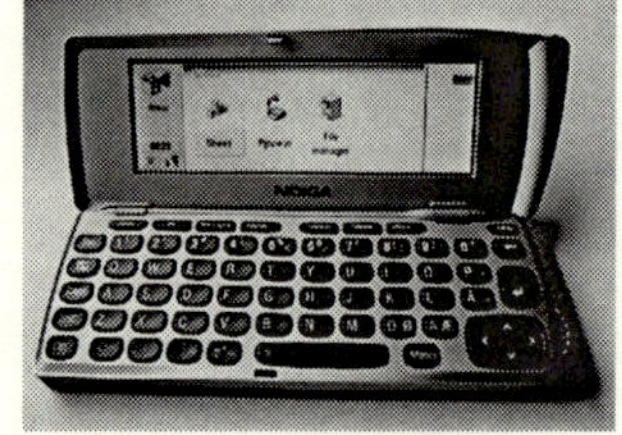

The Nokia 9210 Communicator (Symbian 2000 model smartphone)

In 2000, the touchscreen Ericsson R380 Smartphone was released.[9] It was the first device to use an open operating system, the Symbian OS.[10] It was the first device marketed as a 'smartphone'.[11] It combined the functions of a mobile phone and a personal digital assistant (PDA).[12] In December 1999 the magazine Popular Science appointed the Ericsson R380 Smartphone to one of the most important advances in science and technology.[13] It was a groundbreaking device since it was as small and light as a normal mobile phone.[14] In 2002 it was followed up by P800.[15]

Also in 2000, the Nokia 9210 communicator was introduced, which was the first color screen model from the Nokia Communicator line. It was a true smartphone with an open operating system, the Symbian OS. It was followed by the 9500 Communicator, which also was Nokia's first cameraphone and first Wi-Fi phone. The 9300 Communicator was smaller, and the latest E90 Communicator includes GPS. The Nokia Communicator model is remarkable for also having been the most costly phone model sold by a major brand for almost the full life of the model series, costing easily 20% and sometimes 40% more than the next most expensive smartphone by any major producer.

In 2007 Nokia launched the Nokia N95 which integrated a wide range of multimedia features into a consumer-oriented smartphone: GPS, a 5 megapixel camera with autofocus and LED flash, 3G and Wi-Fi connectivity and TV-out. In the next few years these features would become standard on high-end smartphones. The Nokia 6110 Navigator is a Symbian based dedicated GPS phone introduced in June 2007.

In 2010 Nokia released the Nokia N8 smartphone with a stylus-free capacitive touchscreen, the first device to use the new Symbian^3 OS.[16] It featured a 12 megapixel camera with Xenon flash able to record HD video in 720p, described by Mobile Burn as the best camera in a phone,[17] and satellite navigation that Mobile Choice described as the best on any phone.[18] It also featured a front-facing VGA camera for videoconferencing.

Symbian was the number one smartphone platform by market share from 1996 until 2011 when it dropped to second place behind Google's Android OS. In February 2011, Nokia announced that it would replace Symbian with Windows Phone as the operating system on all of its future smartphones.[19] This transition was completed in October 2011, when Nokia announced its first line of Windows Phone 7.5 smartphones, Lumia 710 and 800.[20]

Palm, Windows, and BlackBerry

In the late 1990s the vast majority of mobile phones had only basic phone features and many people who needed functionality beyond that also carried PDA and/or pager type devices running early versions of operating systems such as Palm OS, BlackBerry OS or Windows CE/Pocket PC.[1] Later versions of these systems started integrating cell phone capabilities with their PDA and messaging features and support of third-party applications. Today, high-end devices running these systems are often branded smartphones.

The HTC Touch Pro2 smartphone (May 2009)

In early 2001, Palm, Inc. introduced the Kyocera 6035, the first smartphone to be deployed in widespread use in the United States. This device combined the features of a personal digital assistant (PDA) with a wireless phone that operated on the Verizon Wireless network. For example, a user could select a name from the PDA contact list, and the device would dial that contact's phone number. The device also supported limited web browsing.[21] The device received a very positive reception from technology publications, but the product line never became widespread outside North America.[22]

In 2001 Microsoft announced its Windows CE Pocket PC OS would be offered as "Microsoft Windows Powered Smartphone 2002."[23] Microsoft originally defined its Windows Smartphone products as lacking a touchscreen and offering a lower screen resolution compared to its sibling Pocket PC devices.

In early 2002 Handspring released the Palm OS Treo smartphone, utilizing a full keyboard that combined wireless web browsing, email, calendar, and contact organizer with mobile third-party applications that could be downloaded or synced with a computer.[24]

In 2002 RIM released their first BlackBerry devices with integrated phone functionality and shifted the positioning of their products from 2-way pagers to email-capable mobile phones. The BlackBerry line evolved into the first smartphone optimized for wireless email use and had achieved a total customer base of about 32 million subscribers by December 2009.[25]

In February 2011 Nokia announced a plan to make Microsoft Windows Phone its main operating system of choice for new Nokia smartphones.[19]

iPhone

The original iPhone (June 2007)

In 2007, Apple Inc. introduced its first iPhone. It was initially costly, priced at $499 for the cheaper of two models on top of a two year contract. The first mobile phone to use a multi-touch interface, the iPhone was notable for its use of a large touchscreen for direct finger input as its main means of interaction, instead of having a stylus, keyboard, and/or keypad, which were the typical input methods for other smartphones at the time. The iPhone featured a web browser that *Ars Technica* then described as "far superior" to anything offered by that of its competitors.[26] Initially lacking the capability to install native applications beyond the ones built-in to its OS, at WWDC in June 2007 Apple announced that the iPhone would support third-party "web 2.0 applications" running in its web browser that share the look and feel of the iPhone interface.[27] As a result of the iPhone's initial inability to install third-party native applications, some reviewers did not consider the originally released device to accurately fit the definition of a smartphone "by conventional terms."[28] A process called jailbreaking emerged quickly to provide unofficial third-party native applications. There are many different functions of the iPhone including a GPS unit, kitchen timer, radio, map book, calendar, notepad, and many other features that replace objects we have to carry around each day.[29]

In July 2008, Apple introduced its second generation iPhone with a lower list price starting at $199 and 3G support. Released with it, Apple also created the App Store, adding the capability for any iPhone or iPod Touch to officially execute additional native applications (both free and paid) installed directly over a Wi-Fi or cellular network, without the more typical process at the time of requiring a PC for installation. Applications could additionally be browsed through and downloaded directly via the iTunes software client on Macintosh and Windows PCs, rather than by searching through multiple sites across the Internet. Featuring over 500 applications at launch,[30] Apple's App Store was immediately very popular,[31] quickly growing to become a huge success.[32] [33]

In June 2010, Apple introduced iOS 4, which included APIs to allow third-party applications to multitask,[34] and the iPhone 4, which included a 960×640 pixel display with a pixel density of 326 pixels per inch (ppi), a 5 megapixel camera with LED flash capable of recording HD video in 720p at 30 frames per second, a front-facing VGA camera for videoconferencing, a 1 GHz processor, and other improvements.[35] In early 2011 the iPhone 4 became available through Verizon Wireless, ending AT&T's exclusivity of the handset in the U.S.,[36] [37] [38] and allowing the handset's 3G connection to be used as a wireless Wi-Fi hotspot for the first time, to up to 5 other devices.[39] Software updates subsequently added this capability to other iPhones running iOS 4.[40] [41]

The iPhone 4S was announced on October 4, 2011, improving upon the iPhone 4 with a dual core A5 processor, an 8 megapixel camera capable of recording 1080p video at 30 frames per second, World phone capability allowing it to work on both GSM & CDMA networks, and the Siri automated voice assistant.[42] On October 10, Apple announced that over one million iPhone 4Ss had been pre-ordered within the first 24 hours of it being on sale, beating the 600,000 device record set by the iPhone 4,[43] [44] despite the iPhone 4S failing to impress some critics at the announcement[45] [46] due to their expectations of an "iPhone 5" with rumored drastic changes compared to the iPhone 4 such as a new case design and larger screen.[47] Along with the iPhone 4S Apple also released iOS 5 and iCloud, untethering iOS devices from Macintosh or Windows PCs for device activation, backup, and synchronization,[48] along with additional new and improved features.[49]

There are about 35 percent of Americans that have some sort of smartphone. This shows that the market is spreading fast and there are also more capabilities for smartphones because of this spread.[50]

Smartphones are also mainly valuable based on the operating system. For example, the iPhone runs on the iOS and other devices run different operating systems which makes the functionality of these systems different.[51]

Android

The Android operating system for smartphones was released in 2008. Android is an open-source platform backed by Google, along with major hardware and software developers (such as Intel, HTC, ARM, Motorola and Samsung, to name a few), that form the Open Handset Alliance.[52] The first phone to use Android was the HTC Dream, branded for distribution by T-Mobile as the G1. The software suite included on the phone consists of integration with Google's proprietary applications, such as Maps, Calendar, and Gmail, and a full HTML web browser. Android supports the execution of native applications and a preemptive multitasking capability (in the form of services). Third-party apps are available via the Android Market (released October 2008), including both free and paid apps.

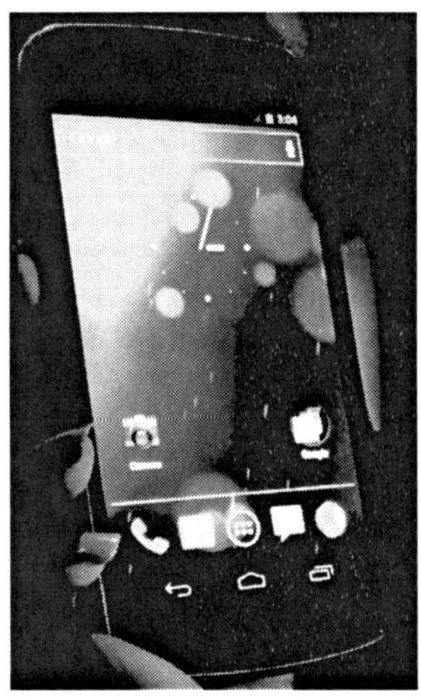
Galaxy Nexus, the latest "Google phone"

In January 2010, Google launched the Nexus One smartphone using its Android OS. Although Android has multi-touch abilities, Google initially removed that feature from the Nexus One,[53] but it was added through a firmware update on February 2, 2010.[54]

Concerning the Xperia Play smartphone, an analyst at CCS Insight said in March 2011 that "Console wars are moving to the mobile platform".[55] In the same month, the HTC EVO 3D was announced by HTC Corporation, which can produce 3D effects with no need for special glasses (autostereoscopy).[56] The HTC EVO 3D was officially released on June 24, 2011.[57]

Bada

The Bada operating system for smartphones was announced by Samsung on 10 November 2009.[58] [59] The first Bada-based phone was the Samsung Wave S8500, released on June 1, 2010,[60] [61] which sold one million handsets in its first 4 weeks on the market.[62]

Samsung shipped 3.5 million phones running Bada in Q1 of 2011.[63] This rose to 4.5 million phones in Q2 of 2011.[64]

Patent licensing and litigation

Recently the number of lawsuits, trade complaints, and countersuits and complaints based on patents and designs in the markets for smartphones, and devices based on smartphone OSes such as Android, has been increasing significantly.

Timeline[65] [66] [67] [68] [69] (initial suits, countersuits, rulings, licence agreements, and other major events in *italics*):

- 2009, Oct 22: *Nokia sues Apple over 10 patents.*[70] [71]
- 2009, Dec 11: *Apple countersues Nokia over 13 patents.*[72]

- 2009, Dec 29: Nokia files a second lawsuit[73] and a U.S. International Trade Commission (ITC) complaint against Apple over 7 more patents.[74]
- 2010, Jan 15: Apple files an ITC complaint against Nokia over 9 patents.[75] [76]
- 2010, Feb 19: Apple drops 4 patents from their countersuit against Nokia that are in their ITC complaint against Nokia.
- 2010, Feb 24: Apple countersues Nokia in Nokia's second lawsuit, over the 9 patents that are in Apple's ITC complaint.
- 2010, Mar 02: *Apple sues HTC over 10 patents and files an ITC complaint against HTC over 10 other patents.*[77] [78] [79] [80] [81]
- 2010, Apr 26: 5 of the patents in Apple's ITC complaint against Nokia are merged into their ITC complaint against HTC.
- 2010, Apr 27: *HTC signs an agreement with Microsoft to licence Microsoft patents in return for royalties on HTC's Android-based devices*[82] [83] (rumored to be $5 per handset).
- 2010, May 7: Nokia files a third lawsuit against Apple over 5 more patents.[84]
- 2010, May 12: *HTC files an ITC complaint against Apple over 5 patents.*[85]
- 2010, May 28: S3 Graphics files an ITC complaint against Apple over 4 patents used in the iPhone, iPod Touch, iPad, and Apple computers.[86]
- 2010, Jun 28: Apple countersues Nokia in Nokia's third lawsuit, over 7 more patents.
- 2010, Jul 06: *HTC countersues Apple over 3 patents.*
- 2010, Jul 21: Nokia drops 1 patent from their ITC complaint against Apple.
- 2010, Aug 12: *Oracle sues Google over 7 patents relating to the use of Java in Android.*[87]
- 2010, Sep 17: Nokia adds 2 more patents to their third lawsuit against Apple.
- 2010, Sep 27: Apple sues Nokia in the UK and Germany over 9 patents.
- 2010, Sep 30: Nokia countersues Apple in Germany over 4 patents.
- 2010, Oct 01: *Microsoft files an ITC complaint and a lawsuit against Motorola over 9 patents.*[88] [89]
- 2010, Oct 06: *Motorola sues Apple over 18 patents, and files an ITC complaint against Apple over 6 of them.*[90]
- 2010, Oct 08: *Motorola files a request for declaratory judgement that they do not infringe 12 Apple patents, and that those patents be declared invalid.*[91] [92]
- 2010, Oct 12: Nokia adds 3 more patents to their countersuit against Apple in Germany.
- 2010, Oct 25: Nokia sues Apple in another German court over 5 patents.
- 2010, Oct 28: Apple drops 4 patents from their ITC complaint against HTC and/or Nokia.
- 2010, Oct 29: *Apple sues Motorola over 6 patents, and files an ITC complaint against Motorola over 3 of them.*[93] [94]
- 2010, Nov 05: HTC drops 1 patent from their ITC complaint against Apple.
- 2010, Nov 09: *Microsoft alleges Motorola has failed to comply with RAND (reasonable and non-discriminatory) licensing obligations.*
- 2010, Nov 10: *Motorola sues Microsoft over 7 patents in one court and 9 patents in another.*
- 2010, Nov 18: *Apple makes counterclaims against Motorola over 6 patents.*
- 2010, Nov 22: *Motorola files an ITC complaint against Microsoft over 5 patents.*
- 2010, Dec 01: *Apple adds the 12 patents to their suit against Motorola that Motorola previously requested declaratory judgement that they do not infringe.*[95]
- 2010, Dec 03: Nokia countersues Apple in the UK over 4 patents, and files a new suit against Apple in the Netherlands over 2 patents.
- 2010, Dec 03: Apple countersues Nokia in Nokia's second German lawsuit, over 1 patent and 2 utility models.
- 2010, Dec 06: Nokia drops 1 patent from their ITC complaint against Apple.
- 2010, Dec 15 and 22: Nokia and Apple take their first German suit/countersuit to the Federal Patent Court of Germany.

- 2010, Dec 23: Motorola files a third lawsuit against Microsoft over 3 patents.
- 2010, Dec 23: Microsoft countersues Motorola over 7 patents.
- 2011, Jan 06: The third Nokia/Apple lawsuit/countersuit is transferred to the location of the first and second ones.
- 2011, Jan 18: Apple seeks to invalidate one Nokia patent in the UK, which it was not yet being sued over.
- 2011, Jan 18: Motorola drops 1 patent from their lawsuits against Microsoft.
- 2011, Jan 19: Microsoft counterclaims against Motorola, asserting 5 patents.
- 2011, Jan 25: Microsoft counterclaims against Motorola, asserting 2 patents.
- 2011, Feb 14: Motorola adds 2 patents to their lawsuits against Microsoft.
- 2011, Feb 22: Apple drops 1 more patent from their ITC complaint against HTC and Nokia.
- 2011, Mar 21: *Microsoft sues Barnes & Noble over the Android operating system in the Nook ebook reader.*[96]
- 2011, Mar 25: *ITC finds that Apple does not infringe on 5 Nokia patents.*
- 2011, Mar 29: Nokia files an ITC complaint against Apple over 7 more patents, and a fourth lawsuit over 6 of those.[97] [98]
- 2011, Apr 15: *Apple sues Samsung for patent and trademark infringement* (7 utility patents, 3 design patents, 3 registered trade dresses, 6 trademarked icons) with its Galaxy line of mobile products, including the *Galaxy S* smartphone and the *Galaxy Tab* tablet.[99] [100]
- 2011, Apr 22: *Samsung sues Apple in South Korea (5 patents), Japan (2 patents), and Germany (3 patents).*[101]
- 2011, Apr 28: *Samsung countersues Apple over 10 patents.*[102]
- 2011, Apr 29: Apple drops 1 more patent from their ITC complaint against HTC.
- 2011, May 18: *Samsung ordered to provide Apple samples of the announced Galaxy S2, Infuse 4G,* and *Infuse 4G LTE* smartphones, as well as the *Galaxy Tab 8.9* and *Galaxy Tab 10.1* tablets as part of Apple's lawsuit against the company.[103] [104]
- 2011, May 18: *Samsung files a court motion for Apple to provide samples of the unannounced iPhone 5 and iPad 3 prototypes.*[105]
- 2011, Jun 14: *Nokia and Apple settle their litigation with Apple agreeing to pay Nokia an undisclosed one-time payment as well as continuing royalties.*[106] [107]
- 2011, Jun 16: *Apple amends its lawsuit against Samsung,* dropping 2 utility patents and 1 design patent, and adding 3 new utility patents plus 4 trade dress applications, *now covering the Samsung Galaxy Tab 10.1*[108]
- 2011, Jun 22: Apple countersues Samsung in South Korea over an unknown number of patents.
- 2011, Jun 22: *Samsung's motion to be provided samples of Apple's unannounced iPhone 5 and iPad 3 prototypes is denied.*[109]
- 2011, Jun 27: General Dynamics Itronix signs an agreement with Microsoft to licence Microsoft patents in return for royalties on General Dynamics Itronix's Android-based devices.[110] [111]
- 2011, Jun 28: Samsung files an ITC complaint and a lawsuit against Apple over 5 patents.
- 2011, Jun 29: Samsung sues Apple in London, UK over an unknown number of patents, and a Samsung lawsuit against Apple in Italy becomes known (details unknown).
- 2011, Jun 29: Velocity Micro signs an agreement with Microsoft to licence Microsoft patents in return for royalties on Velocity Micro's Android-based devices.[111] [112]
- 2011, Jun 30: Samsung converts its countersuit against Apple into counterclaims against Apple's suit, dropping 2 patents but adding 4 more.
- 2011, Jun 30: *A consortium of companies made up of Apple, EMC Corporation, Ericsson, Microsoft, Research In Motion and Sony win against Google*[113] *in an auction of over 6,000 Nortel mobile-related telecommunications patents for $4.5 billion USD.*[114] [115]
- 2011, Jun 30: Onkyo signs an agreement with Microsoft to licence Microsoft patents in return for royalties on Onkyo's Android-based devices.[111] [116]
- 2011, Jul 01: *Apple files for preliminary injunction against 4 Samsung products: Infuse 4G, Galaxy S 4G, Droid Charge,* and *Galaxy Tab 10.1* based on 3 design patents and 1 utility patent.[117]

- 2011, Jul 01: *ITC rules that Apple infringes on 2 patents held by S3 Graphics, while not infringing on 2 others.*[118]
- 2011, Jul 05: *Apple files an ITC complaint against Samsung over 6 smartphones and 2 tablets* infringing 5 utility patents and 2 design patents.
- 2011, Jul 05: Wistron signs an agreement with Microsoft to licence Microsoft patents in return for royalties on Wistron's Android-based devices.[111] [119] [120]
- 2011, Jul 06: *HTC agrees to purchase S3 Graphics* to secure 235 patents for use in its defense against Apple.[121] [122] [123]
- 2011, Jul 06: *Microsoft seeks $15 licensing fees from Samsung for a range of claimed patent violations on every Android device.*[124]
- 2011, Jul 11: Apple files a second ITC complaint against HTC over 5 more patents, and sues HTC over 4 patents from this second ITC complaint that they weren't already suing HTC over.[125] [126]
- 2011, Jul 11-12: Google acquires 1,029 Patents from IBM for an undisclosed amount.[127] [128]
- 2011, Jul 15: *ITC finds HTC infringes on 2 Apple patents.*[129]
- 2011, Jul 29: HTC sues Apple in London, UK over an unknown number of patents.
- 2011, Aug 02: *Apple sues Samsung in Australia over 10 patents, resulting in Samsung delaying the launch and halting advertising of the Samsung Galaxy Tab 10.1 tablet in Australia* to an indefinite date.[130] [131]
- 2011, Aug 09: *A German court issues a preliminary injunction against the Samsung Galaxy Tab 10.1* in Apple's lawsuit against Samsung *which causes its sale to be banned in most of Europe.*[132] [133]
- 2011, Aug 15: *Google announces its intention to buy Motorola Mobility for $12.5 billion USD.* Eighteen of Motorola's patents could potentially be used for defense or countersuits against Apple and Microsoft, and may influence the smartphone war. These patents may change the balance of power, and force the various players to settle their lawsuits.[134] [135]
- 2011, Aug 16: *The Samsung Galaxy Tab 10.1 sales ban in Europe is lifted outside of Germany.*[136] [137]
- 2011, Aug 17: Google acquires 1,023 more patents from IBM for an undisclosed amount (not revealed until 13 Sep 2011).[138]
- 2011, Aug 23: *Microsoft files a complaint with the ITC requesting a ban on several key Motorola smartphones and devices in the USA* based on infringements of 7 patents.[139] [140]
- 2011, Aug 24: *A court in the Netherlands rules that Samsung will be banned from selling the Galaxy S, Galaxy S II and Galaxy Ace in a number of European countries* due to Apple's patent infringement claims.[141]
- 2011, Sep 02: *Apple granted preliminary injunction against Samsung preventing display of the prototype Samsung Galaxy Tab 7.7 tablet at the IFA trade show in Berlin.*[142]
- 2011, Sep 02: *Apple court filings assert that Andy Rubin got inspiration for Android framework while working at Apple* before working at General Magic and Danger, Inc.[143]
- 2011, Sep 07: *HTC countersues Apple using nine patents from Google.* The move is seen as a possible first step for Google giving direct support in lawsuits involving manufacturers using Android.[144] [145] [146] [147]
- 2011, Sep 08: *Acer*[148] *and ViewSonic*[149] *sign patent license agreements with Microsoft* regarding their use of Android on smartphones and tablets.[150] [151]
- 2011, Sep 09: *Apple's preliminary injunction against sales of the Samsung Galaxy Tab 10.1 in Germany is upheld.*[152]
- 2011, Sep 12: Samsung announces a lawsuit against Apple in France that had been filed in July over 3 patents.[153]
- 2011, Sep 12: Apple countersues Samsung in the UK over an unknown number of patents.[154]
- 2011, Sep 13: Google's August 17 acquisition of 1,023 patents from IBM is revealed by the U.S. Patent and Trademark Office.[138] [155]
- 2011, Sep 17: Samsung countersues Apple in Australia over 7 patents.[156]

- 2011, Sep 28: *Samsung signs an agreement with Microsoft to licence Microsoft patents in return for royalties on Samsung's Android-based devices.*[157] [158] [159]
- 2011, Oct 12: *An Australian court issues a preliminary injunction against the Samsung Galaxy Tab 10.1 in Apple's lawsuit against Samsung which prevents its sale in Australia leading up to the 2011 holiday season.*[160]
- 2011, Oct 13: *Quanta signs an agreement with Microsoft to licence Microsoft patents in return for royalties on Quanta's Android and Chrome-based devices.*[161] [162]
- 2011, Oct 13: *Judge in Apple's U.S. lawsuit against Samsung agrees that Samsung's tablets infringe on Apple's patents, but also that the validity of some of the patents might be questionable.*[163]

Screen

Screens on smartphones vary largely in both display size and display resolution. The most common screen sizes range from 2 inches to over 4 inches (measured diagonally). Some 5 inch screen devices exist that run on mobile OSes and have the ability to make phone calls, such as the discontinued Dell Streak and the current Samsung Galaxy Note. Ergonomics arguments have been made that increasing screen sizes start to negatively impact usability.

Common resolutions for smartphone screens vary from 240×320 to 720×1280, with many flagship Android phones at 480×800 or 540×960, the iPhone 4/4S at 640×960 and Galaxy Nexus and HTC Rezound at 720×1280.

Application stores

The introduction of Apple's App Store for the iPhone and iPod Touch in July 2008 popularized manufacturer-hosted online distribution for third-party applications focused on a single platform. Before this, smartphone application distribution was largely dependent on third-party sources providing applications for multiple platforms, such as GetJar, Handango, Handmark, PocketGear, and others.

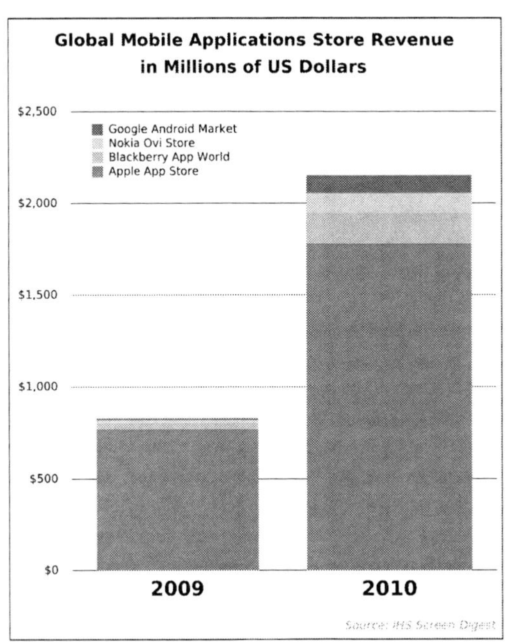

The iPhone's platform is officially restricted to installing apps through the App Store, through "B2B" deployment, and on an "Ad Hoc" basis on up to 100 iPhones.[164] Through jailbreaking it can install apps from other sources. Other platforms may allow application distribution through additional sources outside of their manufacturer-provided app stores, such as third-party app stores and downloads from individual websites.

Following the success of Apple's App Store other smartphone manufacturers quickly launched application stores of their own. Google launched the Android Market in October 2008. RIM launched its app store, BlackBerry App World, in April 2009. Nokia launched its Ovi Store in May 2009. Palm launched its Palm App Catalog for webOS in June 2009. Microsoft launched an application store for Windows Mobile called

Windows Marketplace for Mobile in October 2009, and then a separate Windows Phone Marketplace for Windows Phone in October 2010. Samsung launched Samsung Apps for its Bada based phones in June 2010. Amazon launched its Amazon Appstore for the Google Android operating system in March 2011.

Store	2009 (millions U.S.)	2010 (millions U.S.)[165]
Apple App Store	$769	$1782
Blackberry App World	$36	$165
Nokia Ovi Store	$13	$105
Google Android Market	$11	$102
Total	$828	$2155

The relatively high revenue of U.S. $1782 million in 2010 for Apple's App Store compared to competitor's stores[165] can be attributed to a combination of factors. In large part this can be attributed to having the largest number of apps available and the highest download volume of any mobile app store in 2010, but besides that only 28% of the apps in Apple's App Store were free apps, compared to over 57% in the Android Market. Similarly, Nokia's Ovi Store and the BlackBerry App World both had only 26% of their apps available for free, but both generated higher revenues than the Android Market despite having much lower download volumes.[166]

Malicious software attacks

As smartphone adoption goes up they have increasingly become subject to attacks by malicious software (malware).[167] [168]

Frequently this malware is distributed through application stores that have minimal or no review process for their content.[169] In some cases malware has been hidden in pirated versions of legitimate apps, which are then distributed through 3rd party app stores.[170] [171] Malware risk also comes from what's known as an "update attack," where a legitimate application is later changed to include a malware component, which users then install when they are notified that the app has been updated. Additionally, the ability to acquire software directly from links on the web results in a distribution vector called "malvertizing," where users are directed to click on links, such as on ads that look legitimate, which then open in the device's web browser and cause malware to be downloaded and installed automatically.[172]

Typical smartphone malware leverages platform vulnerabilities that allow it to gain root access on the device in the background. Using this access the malware installs additional software to target communications, location, or other personal identifying information. A common form of malware on mobile phones is the SMS trojan, which sends premium SMS messages, possibly while unknowingly running in the background of a legitimate application. These premium SMS messages run up charges on the owners phone bill which cannot be recovered.

In August 2010, Kaspersky Lab reported detection of the first malicious program for smartphones running on Google's Android operating system, named Trojan-SMS.AndroidOS.FakePlayer.a, an SMS trojan which had already infected a number of devices using that OS.[173] Over the spring of 2011 Android malware increased 76%, according to McAfee.[167] [174] A report from Juniper Global Threat Center notes that malware on the Android platform increased 400% from 2009 to the summer of 2010, and then saw a 472% increase between July and November 2011.[169] The Juniper report indicates that 55% of Android malware acts as spyware, and 44% are SMS trojans.

While there have been and continue to be potential security flaws in iOS,[175] as of at least August 2011 there were no known malware or spyware apps in Apple's App Store, according to security firm Lookout. There are however commercial spyware applications available, outside the App Store, for jailbroken iOS devices.[172] In June 2011 Symantec's 23-page report "A Window Into Mobile Device Security" characterized (non-jailbroken) devices running iOS as having "full protection" against malware attacks.[176]

Symbian and older versions Windows Mobile have had to contend with a degree of malware in the past, but as legacy systems it is believed that the people who previously targeted them have shifted their focus to Android.[169] There were also a few Palm OS viruses.

The only mobile platform other than Apple's iOS without reports of malware so far is HP's (formerly Palm's) webOS, but this may be explained by its relatively low adoption rate.[174]

The best way to reduce a device's vulnerability to malware attacks is to install the most recent versions of operating systems which include security patches. This can be complicated by long delays[177] in software updates for many devices which have had their software modified with custom "skins," services, or promotional on-deck apps by their manufacturer or mobile carrier.[172] In some cases a device may no longer be receiving updates from its manufacturer or carrier, leaving it vulnerable to exploits that have been patched in an OS version that's more recent than the device's last supported one.

Market share

Smartphone market share

For several years, demand for advanced mobile devices boasting powerful processors and graphics processing units, abundant storage (flash memory) for applications and media files, high-resolution screens with multi-touch capability, and open operating systems has outpaced the rest of the mobile phone market.[178]

According to an early 2010 study by ComScore, over 45.5 million people in the United States owned smartphones out of 234 million total subscribers.[179] Despite the large increase in smartphone sales in the last few years, smartphone shipments only made up 20% of total handset shipments as of the first half of 2010.[180]

According to Gartner in their report dated November 2010, total smartphone sales doubled in one year and now smartphones represent 19.3 percent of total mobile phone sales.[181] Smartphone sales increased in 2010 by 72.1 percent from the prior year, whereas sales for all mobile phones only increased by 32%.[182] [183]

According to an Olswang report in early 2011, the rate of smartphone adoption is accelerating: as of March 2011 22% of UK consumers had a smartphone, with this percentage rising to 31% amongst 24- to 35-year-olds.[184]

In March 2011, Berg Insight reported data that showed global smartphone shipments increased 74% from 2009 to 2010.[185]

A survey of mobile users in the United States by Nielsen in Q3, 2011 reports that smartphone ownership has reached 43% of all U.S. mobile subscribers, with the vast majority of users under the age of 44 owning one. In the 25-34 age range smartphone ownership is reported to be at 62%.[186] NPD Group reports that the share of handset sales that were smartphones in Q3, 2011 reached 59% for consumers 18 and over in the U.S.[187]

In profit share worldwide smartphones now far exceed the share of non-smartphones. According to a November 2011 research note from Canaccord Genuity, Apple Inc. holds 52% of the total mobile industry's operating profits, while only holding 4.2% of the global handset market. HTC and RIM similarly only make smartphones and their wordwide profit shares are at 9% and 7%, respectively. Samsung, in second place after Apple at 29%, makes both smartphones and feature phones and doesn't report a breakdown separating their profits between the two kinds of devices, but it can be intuited that a significant portion of that profit comes from their flagship smartphone devices.[188]

Up to the end of November 2011, camera-equipped smartphones took 27 percent of photos, a significant increase from 17 percent last year. Due to the fact that we carry smartphones with us all the time, smartphones have replaced some functions of Point-and-shoot cameras, except the cameras with big optical zoom such as 10x.[189]

Operating system market shares

2010 saw the rapid rise of the Google Android operating system from 4 percent of new deployments in 2009 to 33 percent at the beginning of 2011 making it share the top position with the since long dominating Symbian OS. The smaller rivals include US popular Blackberry OS, iOS, Samsung's recently introduced Bada, HP's heir of Palm webOS and the Microsoft Windows Phone OS which is now supported by Nokia.

Quantity market shares by Gartner (in one year) (new sales)

Smartphone OS	Percent
Android 2009	3.9%
Android 2010	22.7%
Symbian 2009	46.9%
Symbian 2010	37.6%
RIM 2009	19.9%
RIM 2010	16.0%
iOS 2009	14.4%
iOS 2010	15.7%
Windows 2009	8.7%
Windows 2010	4.2%
Other 2009	6.1%
Other 2010	3.8%

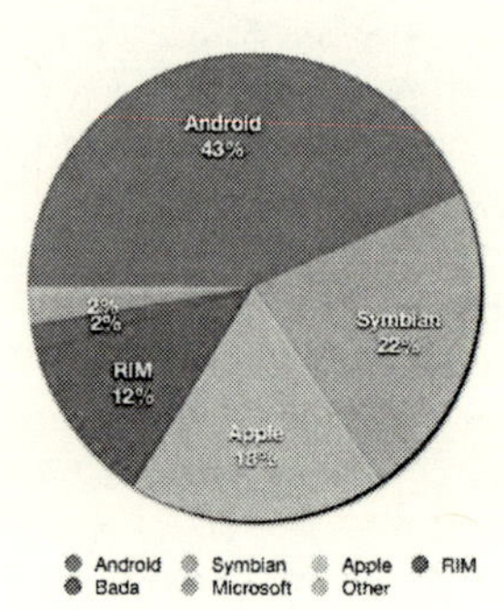

Share of worldwide 2011 Q2 smartphone sales to end users by operating system, according to Gartner.[190]

Over late 2009 and 2010 Android's smartphone operating system market share increased very rapidly.[181] In the fourth quarter of 2010, Android surpassed Symbian as the most common operating system in smartphones, with 32.9 million units sold versus 31.0 million. Android-equipped phones sold seven times more than in the prior year.[191] According to Canalys, Google's Android operating system, which is offered to phone makers for free, has raced to the top past operating systems by Nokia, Apple, RIM, and Microsoft. In Q1 2011 Google's Android market share was 35 percent, increasing significantly from 10 percent the previous year, while Nokia's Symbian dropped to 26 percent from 46 percent over the same time period.[192] In the UK, which currently has one of the highest penetrations of smartphones in the World, Android achieved 50% market share in October 2011.[193]

Historical market share

Figures in millions.

Year	Android (Google)	Blackberry (RIM)	IPhone (Apple)	Linux	Palm/WebOS (Palm/HP)	Symbian (Nokia)	Windows Mobile/Phone (Microsoft)
2007[194]		11.77	3.3	11.76	1.76	77.68	14.7
2008[195]		23.15	11.42	11.26	2.51	72.93	16.5
2009[196]	6.8	34.35	24.89	8.13	1.19	80.88	15.03
2010[197]	67.22	47.45	46.6			111.58	12.38

Enterprise share by operating system

In a worldwide study of 2,300 workers at 1,100 businesses by iPass it was reported that Apple's iPhones have displaced RIM's BlackBerry devices in enterprise adoption in 2011.[198] The share for iPhones increased to 45% from 31.1% in 2010, while the Blackberry share dropped to 32.2% from 34.5% in the previous year. Android phones also increased in share, to 21.3% from 11.3% in 2010, exceeding Symbian for the first time, which dropped to 7.4% from 12.4%. Windows Mobile and all other smartphone OSes also dropped in 2011 compared to 2010.[199]

Customer loyalty by operating system

According to a survey of more than 6,000 smartphone users through 2010 by mobile analytics firm Zokem, the top five loyalty scores for smartphone platforms are the iPhone at 73%, followed by Google's Android at 40%, Samsung's Bada at 33%, RIM's BlackBerry at 30%, and Symbian S60 at 23%. Windows Mobile and Palm follow at 10% each. Customer loyalty gauges the likelihood that the user of a smartphone platform whose contract has expired or who has broken or lost their phone will repurchase another one that uses that same platform.[200] [201]

Manufacturer market shares

From the launch of their Communicator model in 1996 until 2011 Nokia was dominant in the smartphone market, though has more recently been joined by other competitors in the market. Based on a report by Strategy Analytics, Samsung overtook Nokia in smartphone shipments with an estimated 27.8 million units shipped in Q3 2011[202] (Samsung does not publicly disclose the numbers of their smartphone shipments and sales).

Quantity market shares by Strategy Analytics (note that Gartner instead shows Nokia ahead on Apple sales based on Symbian sales only[203] (new sales)

Manufacturer	Percent
Apple Q2 2010	13.5%
Apple Q2 2011	18.5%
Samsung Q2 2010	5.0%
Samsung Q2 2011	17.5%
Nokia Q2 2010	38.1%
Nokia Q2 2011	15.2%
Others Q2 2010	43.4%
Others Q2 2011	48.9%

Market share among smartphone manufacturers does not resemble smartphone OS market share numbers due to the differences between the two major smartphone OS sales models: single manufacturer and licensed. Apple's iPhone, Nokia's Symbian, and RIM's BlackBerry smartphones are currently only available from single manufacturers.

Google's Android OS and Microsoft's mobile OSes are platforms that are licensed and used by a variety of manufacturers. As a result, manufacturers of smartphones using licensed OSes all split the total market share of that OS between them, while the total share for a single-manufacturer OS is held by that manufacturer alone.

Note that Nokia's Symbian OS was previously available from several manufacturers under a licensed model, then later predominantly only by Nokia itself more like a single manufacturer model.

Samsung smartphones use a diverse portfolio of operating systems, including their own Bada operating system along with Android and Windows Mobile.[204]

Apple surpassed Nokia worldwide by revenue and profit for the first time in Q2 2011 (though not in market share), with Apple's profit share of the total worldwide smartphone market increasing to 66.3% while Nokia reported a loss.[205]

Between Q2 2010 and Q2 2011 Nokia's worldwide Symbian smartphone sales dropped significantly from 38.1 percent to 15.2 percent, while Samsung smartphone sales increased significantly worldwide from 5% to 17.5%.[206] As of Q1 2011, Nokia had already announced plans to switch to Windows Phone.

Smartphone Customer Satisfaction by J.D. Power and Associates

Manufacturer	Score
Apple 2010	810
Apple 2011	838
HTC 2010	727
HTC 2011	801
Industry Average 2010	753
Industry Average 2011	788
Samsung 2010	724
Samsung 2011	777
Motorola 2010	N/A
Motorola 2011	775
RIM 2010	741
RIM 2011	762
LG 2010	N/A
LG 2011	760
HP/Palm 2010	712
HP/Palm 2011	733
Nokia 2010	720
Nokia 2011	721

Rankings are based on a possible top score of 1000

Nokia still remains the number one company in the worldwide mobile phone market with sales for Q2 2011 of 88.5 million when including feature phone platforms such as S40, compared with 16.7 million smartphones running Symbian.[207]

According to Nielsen in July 2011, in the United States Apple is the top smartphone manufacturer at 28% of the market, with RIM at 20%. Google Android has 39% of the U.S. market as a whole, but this is split between HTC at 14%, Motorola at 11%, Samsung at 8%, and other remaining manufacturers at 6%. HTC's total share of the U.S.

smartphone market actually ties RIM at 20%, since sales of their smartphones running Microsoft's mobile operating systems account for 6% of the total market. Samsung similarly gains 2% of overall U.S. market share due to their sales of Microsoft OS-based smartphones. In contrast to the worldwide market, Nokia's share of U.S. smartphone sales is very small, at only 2%.[208] [209] Nielsen's Q3, 2011 survey of mobile users maintains Apple as the top U.S. smartphone maker with a continued 28% of the market, with RIM dropping from 20% to 18%.[186] While Google Android increased in total operating system share from 39% to 43% of the U.S. market, it remains fragmented amongst many different manufacturers. Over the same quarter Microsoft managed a modest gain from 6% to 7% total U.S. smartphone OS share.

Checks with U.S. carriers by technology analyst firm Canaccord Genuity in April and August 2011 have found that Apple's iPhone 4 has consistently been the top selling device at AT&T and Verizon. In addition, the second most popular spot at AT&T has been maintained by the iPhone 3GS, which was originally released in 2009 (and has never been sold on Verizon). In August 2011 the most popular smartphones on Sprint and T-Mobile in the U.S. were the HTC EVO 3D 4G and HTC Sensation, respectively. The other second most popular smartphones were the Samsung Charge 4G on Verizon, the Motorola Photon 4G on Sprint, and the HTC myTouch 4G Slide on T-Mobile.[210] [211] NPD Group reported that in Q3, 2011 the overall top 5 smartphones by sales across all carriers in the U.S. were, in order: the iPhone 4, iPhone 3GS, HTC EVO 4G, Motorola Droid 3, and Samsung Intensity II.[187]

Currently the vast majority of smartphones are manufactured in China, Taiwan and Mexico, for companies based in the U.S. (Apple, HP, Motorola), South Korea (LG, Samsung), Canada (RIM), Finland (Nokia), Taiwan (HTC) and the U.K. (Sony Ericsson).

In Q3 2011, Samsung became the world's number one smartphone vendor, selling 24 million smartphones. Nokia's Symbian platform remained in second place with 19.5 million Symbian smartphones, and Apple fell to third place, with 17 million phones. Android had 52.5% of the market, with 60.5 million Android phones being sold. In the US, HTC became the largest smartphone vendor.[212]

Customer satisfaction by manufacturer

According to global marketing information services firm J.D. Power and Associates smartphones from Apple Inc. have been consistently[213] ranking highest in customer satisfaction,[214] [215] with a late 2011 score of 838 out of 1000. Based on the responses to their most recent survey of 6,898 smartphone users, Apple was followed in ranking by HTC (801), Samsung (777), Motorola (775), RIM (762), LG (760), Palm (733), and Nokia (721).[216] [217] [218] [219]

Open-source development

The open-source culture has penetrated the smartphone market in several ways. There have been attempts to open source both hardware and software of smartphones.

In February 2010 Nokia made Symbian open source. Thus, most commercial smartphones were based on open-source operating systems. These include those based on Linux, such as Google's Android, Nokia's Maemo, Hewlett-Packard's webOS, and those based on BSD, such as the Darwin-based Apple iOS. Maemo was later merged with Intel's project Moblin to form MeeGo.[220] [221]

Popular services

Location-based check-in services

According to a ComScore report released on May 12, 2011, nearly one in five smartphone users are tapping into check-in services like Foursquare and Gowalla. A total of 16.7 million mobile-phone subscribers used location-based services on their phones in March 2011.[222]

Second screen

The smartphones have introduced a new way of watching television.[223] The second screen is a consequence of the media multitasking which is exploding.[224]

See also

- Camera phone and videophone
- Comparison of smartphones
- List of digital distribution platforms for mobile devices
- Mobile broadband connectivity
- Mobile Internet device (MID) and personal digital assistant (PDA)
- Mobile operating system
- Second screen

References

[1] "Smartphone" (http://www.phonescoop.com/glossary/term.php?gid=131). *Phone Scoop*. . Retrieved 2011-12-15.

[2] "Feature Phone" (http://www.phonescoop.com/glossary/term.php?gid=310). *Phone Scoop*. . Retrieved 2011-12-15.

[3] Andrew Nusca (20 August 2009). "Smartphone vs. feature phone arms race heats up; which did you buy?" (http://www.zdnet.com/blog/gadgetreviews/smartphone-vs-feature-phone-arms-race-heats-up-which-did-you-buy/6836). ZDNet. . Retrieved 2011-12-15.

[4] "Smartphone definition from PC Magazine Encyclopedia" (http://www.pcmag.com/encyclopedia_term/0,2542,t=Smartphone&i=51537,00.asp). *PC Magazine*. . Retrieved 2011-12-15.

[5] Schneidawind, J: "Big Blue unveiling", *USA Today*, November 23, 1992, p. 2B.

[6] "Ericsson GS88 Preview" (http://pws.prserv.net/Eri_no_moto/GS88_Preview.htm). *Eri-no-moto*. . Retrieved 2011-12-15.

[7] "History" (http://www.stockholmsmartphone.org/history/). Stockholm Smartphone. . Retrieved 2011-12-15.

[8] "Ericsson GS88 box" (http://www.stockholmsmartphone.org/wp-content/uploads/penelope-box.jpg). . Retrieved 2011-12-15.

[9] "PDA Review: Ericsson R380 Smartphone" (http://www.geek.com/hwswrev/pda/ericr380/). *Geek.com*. . Retrieved 2011-12-15.

[10] "Symbian Device – The OS Evolution" (http://www.i-symbian.com/wp-content/uploads/2009/11/Symbian_Evolution.pdf) (PDF). *Independent Symbian Blog*. . Retrieved 2011-12-15.

[11] "Ericsson Introduces The New R380e" (http://www.mobilemag.com/2001/09/25/ericsson-introduces-the-new-r380e). *Mobile Magazine*. . Retrieved 2011-12-15.

[12] "Ericsson R380 World, Review & Rating" (http://www.pcmag.com/article2/0,2817,40827,00.asp). *PCMag.com*. . Retrieved 2011-12-15.

[13] *Popular Science, December 1999* (http://books.google.com/books?id=8qSgh_Q-YOkC). *Google Books*. . Retrieved 2011-12-15.

[14] "Ericsson R380 PDA & Phone" (http://www.cellular.co.za/ericsson_r380.htm). *CellularOnline*. . Retrieved 2011-12-15.

[15] "Sony Ericsson P800" (http://www.allaboutsymbian.com/features/item/Sony_Ericsson_P800.php). *All About Symbian*. . Retrieved 2011-12-15.

[16] "Nokia N8 smartphone official site" (http://www.nokia.co.uk/gb-en/products/phone/n8-00/). . Retrieved 2011-12-15.

[17] "Nokia N8 review - best camera phone ever" (http://www.mobileburn.com/review.jsp?Id=11093). Mobileburn.com. 2010-10-05. . Retrieved 2011-12-15.

[18] "Mobile Choice Award: Best Sat-Nav" (http://www.mobilechoiceuk.com/News/mobile+choice+award:+Best+Sat-Nav/5274). Mobile Choice. 22 October 2010. . Retrieved 2011-12-15.

[19] "Nokia, Microsoft in pact to rival Apple, Google - Technology & Science" (http://www.cbc.ca/news/technology/story/2011/02/11/nokia-microsoft-smart-phone-apple-google.html). Associated Press. CBC.ca. 2011-02-11. . Retrieved 2011-12-15.

[20] Velazco, Chris (26 October 2011). "Nokia Debuts Their First Windows Phones: The Lumia 800 and Lumia 710" (http://techcrunch.com/2011/10/26/nokia-debuts-lumia-710-and-lumia-800/). *TechCruch*. . Retrieved 26 October 2011.

[21] "Kyocera QCP 6035 Smartphone Review" (http://www.palminfocenter.com/view_story.asp?ID=1707). Palminfocenter.com. 2001-03-16. . Retrieved 2011-09-07.

[22] Segan, Sascha (2010-03-23). "Kyocera Launches First Smartphone In Years | News & Opinion" (http://www.pcmag.com/article2/0,2817,2361664,00.asp). PCmag.com. . Retrieved 2011-09-07.

[23] "Better Living through Software: Microsoft Advances for the Home Highlighted at Consumer Electronics Show 2002: Microsoft showcases new and improved products -- including "Freestyle," "Mira" and Ultimate TV at International CES 2002" (http://www.microsoft.com/presspass/features/2002/Jan02/01-08msces.mspx). Microsoft.com. 2002-01-08. . Retrieved 2011-09-07.

[24] Stephen H. Wildstrom (November 30, 2001). "Handspring's Breakthrough Hybrid" (http://www.businessweek.com/bwdaily/dnflash/nov2001/nf20011129_0157.htm). Businessweek.com. . Retrieved 2011-12-15.

[25] Kevin McLaughlin (December 17, 2009). "BlackBerry Users Call For RIM To Rethink Service" (http://www.crn.com/news/client-devices/222002587/blackberry-users-call-for-rim-to-rethink-service.htm). CRN.com. . Retrieved 2011-12-15.

[26] "iPhone in depth: the Ars review" (http://arstechnica.com/apple/reviews/2007/07/iphone-review.ars/6). *ArsTechnica*. Condé Nast. 9 July 2007. p. 6. . Retrieved 3 August 2010.

[27] "iPhone to Support Third-Party Web 2.0 Applications" (http://www.apple.com/pr/library/2007/06/11iphone.html). *Press Release*. Apple Inc.. June 11, 2007. . Retrieved December 15, 2008.

[28] "The iPhone is not a smartphone" (http://www.engadget.com/2007/01/09/the-iphone-is-not-a-smartphone/). Engadget.com. 9 January 2007. . Retrieved 11 July 2010.

[29] "Smartphones Can Replace These Everyday Items" (http://simpleorganizedlife.com/smartphones-can-replace-these-everyday-items/). Simple. Organized. Life.. October 10, 2011. . Retrieved 2011-12-15.

[30] "iPhone 3G on Sale Tomorrow" (http://www.apple.com/pr/library/2008/07/10iphone.html). *Press Release*. Apple Inc.. 2008-07-10. . Retrieved 2009-01-17.

[31] "iPhone App Store Downloads Top 10 Million in First Weekend" (http://www.apple.com/pr/library/2008/07/14iPhone-App-Store-Downloads-Top-10-Million-in-First-Weekend.html). *Press Release*. Apple Inc.. July 14, 2008. . Retrieved 2011-12-15

[32] "Apple's App Store Downloads Top 1.5 Billion in First Year" (http://www.apple.com/pr/library/2009/07/14Apples-App-Store-Downloads-Top-1-5-Billion-in-First-Year.html). *Press Release*. Apple Inc.. July 14, 2009. . Retrieved 2011-12-15.

[33] "Apple's App Store Downloads Top 15 Billion" (http://www.apple.com/pr/library/2011/07/07Apples-App-Store-Downloads-Top-15-Billion.html). *Press Release*. Apple Inc.. July 7, 2011. . Retrieved 2011-12-15.

[34] "Apple now accepting iOS 4 apps, multitasking ahoy" (http://www.engadget.com/2010/06/11/apple-now-accepting-ios-4-apps-multitasking-ahoy/). Engadget.com. 11 June 2010. . Retrieved 2011-12-15.

[35] "Apple Presents iPhone 4" (http://www.apple.com/pr/library/2010/06/07Apple-Presents-iPhone-4.html). *Press Release*. Apple Inc.. 2010-06-07. . Retrieved 2011-07-05.

[36] "Liveblog: The Verizon iPhone" (http://voices.washingtonpost.com/fasterforward/2011/01/liveblog_the_verizon_iphone.html). *The Washington Post*. .

[37] Memmott, Mark (2011-01-11). "It's Official: Verizon Has The iPhone 4 : The Two-Way" (http://www.npr.org/blogs/thetwo-way/2011/01/11/132833078/its-official-verizon-has-iphone-4). NPR. . Retrieved 2011-09-07.

[38] Raice, Shayndi (January 12, 2011). "Verizon Unwraps iPhone" (http://online.wsj.com/article/SB10001424052748703791904576075681886276172.html?mod=googlenews_wsj). *The Wall Street Journal*. .

[39] Glenn Fleishman (2011-02-22). "Using the Personal Hotspot on your Verizon iPhone" (http://www.macworld.com/article/158058/2011/02/personal_hotspot_verizon.html). Macworld.com. . Retrieved 2011-03-12.

[40] "iOS 4.3 Software Update" (http://www.apple.com/ios/). Apple Inc.. . Retrieved 2011-03-12.

[41] Dan Moren (2011-03-11). "Hands on with iOS 4.3" (http://www.macworld.com/article/158483/2011/03/firstlook_43.html). Macworld.com. . Retrieved 2011-03-12.

[42] "Apple Launches iPhone 4S, iOS 5 & iCloud" (http://www.apple.com/pr/library/2011/10/04Apple-Launches-iPhone-4S-iOS-5-iCloud.html). Apple Inc.. October 4, 2011. . Retrieved 2011-12-15.

[43] "iPhone 4S Pre-Orders Top One Million in First 24 Hours" (http://www.apple.com/pr/library/2011/10/10iPhone-4S-Pre-Orders-Top-One-Million-in-First-24-Hours.html). Apple. . Retrieved 10 October 2011.

[44] Anderson, Ash. "iPhone 4S Sells 1 Million in Under 24 Hours" (http://www.keynoodle.com/iphone-4s-sells-1-million-in-under-24-hours/). *KeyNoodle*. . Retrieved 2011-12-15.

[45] "Apple's fall from grace" (http://www.apple.com/pr/library/2011/10/04Apple-Launches-iPhone-4S-iOS-5-iCloud.html). BGR.com. October 5, 2011. . Retrieved 2011-12-15.

[46] Michael E. Cohen (October 13, 2011). "iPhone 4S: A Very Palpable Hit" (http://tidbits.com/article/12554). tidbits.com. . Retrieved 2011-12-15.

[47] David Pogue (October 11, 2011). "New iPhone Conceals Sheer Magic" (http://www.nytimes.com/2011/10/12/technology/personaltech/iphone-4s-conceals-sheer-magic-pogue.html?_r=2&pagewanted=all). The New York Times. . Retrieved 2011-12-15.

[48] "Press Info - Apple to Launch iCloud on October 12" (http://www.apple.com/pr/library/2011/10/04Apple-to-Launch-iCloud-on-October-12.html). Apple. 2011-10-04. . Retrieved 2012-01-05.

[49] "Press Info - New Version of iOS Includes Notification Center, iMessage, Newsstand, Twitter Integration Among 200 New Features" (http://www.apple.com/pr/library/2011/06/06New-Version-of-iOS-Includes-Notification-Center-iMessage-Newsstand-Twitter-Integration-Among-200-New-Features.html). Apple.

2011-06-06. . Retrieved 2012-01-05.

[50] Michael B. Farrell (November 12, 2011). "No cash, card? No problem: Paying by smartphone is an emerging trend" (http://articles.boston.com/2011-11-12/business/30391540_1_smartphone-mobile-commerce-card-readers). boston.com. . Retrieved 2011-12-15.

[51] 13 November 2011. (http://cellphones.about.com/od/smartphonebasics/a/what_is+_smart.htm)

[52] "Alliance Members" (http://www.openhandsetalliance.com/oha_members.html). *Open Handset Alliance*. . Retrieved 16 January 2011.

[53] "Weighing Nexus over iPhone – a practical review for the everyday user!" (http://techietrick.blogspot.com/2010/01/weighing-nexus-over-iphone-practical.html). Techietrick.blogspot.com. 2010-01-06. . Retrieved 2011-09-07.

[54] "Nexus One gets a software update, enables multitouch (updated with video!)" (http://www.engadget.com/2010/02/02/nexus-one-gets-a-software-update-enables-multitouch). Engadget.com. . Retrieved 2011-09-07.

[55] Tarmo Virki (2011-02-14). "Sony takes gaming console war to phones" (http://in.reuters.com/article/2011/02/13/idINIndia-54865020110213). Reuters. .

[56] "HTC EVO 3D" (http://web.archive.org/web/20110511105547/http://www.htc.com/www/product/evo3d/overview.html). htc.com. Archived from the original (http://www.htc.com/www/product/evo3d/overview.html) on 2011-05-11. . Retrieved 2011-12-15.

[57] Savov, Vlad. "HTC Evo 3D Launches June 24th" (http://www.engadget.com/2011/06/06/htc-evo-3d-launches-on-june-24th-for-200-joined-by-evo-view-4g/). Engadget.com. . Retrieved 7 June 2011.

[58] Ed Hansberry (11 November 2009). "Samsung Bailing on Windows Mobile" (http://www.informationweek.com/blog/main/archives/2009/11/samsung_bailing.html). *InformationWeek*. .

[59] "Samsung to Discard Windows Phone" (http://www.telecomskorea.com/market-8281.html). *Telecoms Korea*. 9 November 2009. .

[60] "Samsung Wave, first Bada smartphone hits the market" (http://www.bada.com/samsung-wave-first-bada-smartphone-hits-the-market/). *Bada*. 24 May 2010. . Retrieved 3 February 2011.

[61] "BadaWave" (http://badawave.com/). BadaWave. . Retrieved 2012-01-05.

[62] "Samsung Waves away a million" (http://www.theinquirer.net/inquirer/news/1722287/samsung-waves-away-million). The Inquirer. 13 July 2010. .

[63] "Android increases smart phone market leadership with 35% share" (http://www.canalys.com/newsroom/android-increases-smart-phone-market-leadership-35-share). Canalys.com. 2011-05-04. . Retrieved 2011-09-07.

[64] "Samsung Bada shipments up 355% to 4.5 million units in Q2 2011 | asymco news | PG.Biz" (http://www.pocketgamer.co.uk/r/PG.Biz/asymco+news/news.asp?c=32049). Pocket Gamer. . Retrieved 2011-09-07.

[65] Florian Mueller. "Apple vs Android 10.12.02" (http://www.scribd.com/doc/44759893/Apple-vs-Android-10-12-02). . Retrieved 2011-09-08.

[66] Florian Mueller. "NokiaVsApple_11.03.31.100" (http://www.scribd.com/doc/52195210/NokiaVsApple-11-03-31-100). . Retrieved 2011-09-08.

[67] Florian Mueller. "Microsoft vs Motorola 11.04.09" (http://www.scribd.com/doc/52754079/Microsoft-vs-Motorola-11-04-09). . Retrieved 2011-09-08.

[68] Florian Mueller. "AppleVsSamsung_11.07.05" (http://www.scribd.com/doc/59474521/AppleVsSamsung-11-07-05). . Retrieved 2011-09-08.

[69] Florian Mueller. "AppleVsHTCandS3_11.07.29" (http://www.scribd.com/doc/61387056/AppleVsHTCandS3-11-07-29). . Retrieved 2011-09-08.

[70] "Nokia sues Apple over iPhone's use of patented wireless standards" (http://www.appleinsider.com/articles/09/10/22/nokia_sues_apple_over_iphones_use_of_patented_wireless_standards.html). Appleinsider.com. 2009-10-22. . Retrieved 2012-01-05.

[71] "Nokia vs. Apple: the in-depth analysis" (http://www.engadget.com/2009/10/29/nokia-vs-apple-the-in-depth-analysis/). Engadget.com. . Retrieved 2012-01-05.

[72] "Apple countersues Nokia for infringing 13 patents" (http://www.engadget.com/2009/12/11/apple-countersues-nokia-for-infringing-13-patents/). Engadget.com. . Retrieved 2012-01-05.

[73] Foresman, Chris (2010-01-04). "Nokia adds additional lawsuit in patent catfight with Apple" (http://arstechnica.com/apple/news/2010/01/nokia-adds-additional-lawsuit-in-patent-catfight-with-apple.ars). Arstechnica.com. . Retrieved 2012-01-05.

[74] Foresman, Chris (2009-12-29). "Nokia hurls new salvo in spat with Apple, complains to ITC" (http://arstechnica.com/apple/news/2009/12/nokia-hurls-new-salvo-in-spat-with-apple-complains-to-itc.ars). Arstechnica.com. . Retrieved 2012-01-05.

[75] Decker, Susan (2010-01-16). "Apple Files New Trade Complaint With Against Nokia" (http://www.bloomberg.com/apps/news?pid=newsarchive&sid=ao_5HVbD_IRM). Bloomberg.com. . Retrieved 2012-01-05.

[76] Cheng, Jacqui (2010-01-18). "Apple wants Nokia's US imports blocked" (http://arstechnica.com/apple/news/2010/01/apple-continues-nokia-patent-spat-with-its-own-itc-complaint.ars). Arstechnica.com. . Retrieved 2012-01-05.

[77] "Apple vs HTC: a patent breakdown" (http://www.engadget.com/2010/03/02/apple-vs-htc-a-patent-breakdown/). Engadget.com. 2010-03-02. . Retrieved 2012-01-05.

[78] "Google backs HTC in what could be 'long and bloody battle' with Apple" (http://www.appleinsider.com/articles/10/03/03/google_backs_htc_in_what_could_be_long_and_bloody_battle_with_apple.html). Appleinsider.com. 2010-03-03. . Retrieved 2012-01-05.

[79] Bilton, Nick (2010-03-02). "What Apple vs. HTC Could Mean" (http://bits.blogs.nytimes.com/2010/03/02/what-apple-vs-htc-could-mean/). Bits.blogs.nytimes.com. . Retrieved 2012-01-05.

[80] "HTC says it uses own technology, not Apple's" (http://www.macworld.com/article/146819/2010/03/htc_defense.html). Macworld.com. . Retrieved 2012-01-05.

[81] Foresman, Chris (2010-03-09). "HTC lawsuit came after warning by Apple to handset makers" (http://arstechnica.com/apple/news/2010/03/htc-lawsuit-came-after-warning-by-apple-to-handset-makers.ars). Arstechnica.com. . Retrieved 2012-01-05.
[82] "Microsoft Announces Patent Agreement With HTC" (http://www.microsoft.com/Presspass/press/2010/apr10/04-27MSHTCPR.mspx). Microsoft.com. 2010-04-27. . Retrieved 2012-01-05.
[83] "HTC licenses Microsoft's mobile patents for Android phones – First Take" (http://gartenberg.wordpress.com/2010/04/28/htc-licenses-microsofts-mobile-patents-for-android-phones-first-take/). Gartenberg.wordpress.com. 2010-04-28. . Retrieved 2012-01-05.
[84] Johnston, Casey (2010-05-07). "Nokia applies patent thumbscrews to Apple's iPad" (http://arstechnica.com/apple/news/2010/05/nokia-applies-patent-thumbscrews-to-apples-ipad.ars). Arstechnica.com. . Retrieved 2012-01-05.
[85] "HTC countersues Apple, claims infringement of five patents" (http://www.appleinsider.com/articles/10/05/12/htc_countersues_apple_claims_infringement_of_five_patents.html). Appleinsider.com. 2010-05-12. . Retrieved 2012-01-05.
[86] "S3 Graphics Files New 337 Complaint Regarding Certain Electronic Devices With Image Processing Systems" (http://www.itcblog.com/20100602/s3-graphics-files-new-337-complaint-regarding-certain-electronic-devices-with-image-processing-systems/). Itcblog.com. 2010-06-02. . Retrieved 2012-01-05.
[87] August 12, 2010 (2010-08-12). "Oracle sues Google over Android" (http://venturebeat.com/2010/08/12/oracle-sues-google-over-android/). Venturebeat.com. . Retrieved 2012-01-05.
[88] Protalinski, Emil (2010-10-02). "Microsoft sues Motorola, citing Android patent infringement" (http://arstechnica.com/microsoft/news/2010/10/microsoft-sues-motorola-citing-android-patent-infringement.ars). Arstechnica.com. . Retrieved 2012-01-05.
[89] "Microsoft Files Patent Infringement Action Against Motorola" (http://www.microsoft.com/presspass/press/2010/oct10/10-01statement.mspx). Microsoft.com. 2010-10-01. . Retrieved 2012-01-05.
[90] Foresman, Chris (2010-10-06). "Motorola asks ITC, two federal courts to throw book at Apple" (http://arstechnica.com/apple/news/2010/10/motorola-asks-itc-two-federal-courts-to-throw-book-at-apple.ars). Arstechnica.com. . Retrieved 2012-01-05.
[91] "Motorola Moves Offensively with a New Lawsuit against Apple" (http://www.patentlyapple.com/patently-apple/2010/10/motorola-moves-offensively-with-a-new-lawsuit-against-apple.html). Patentlyapple.com. 2010-10-15. . Retrieved 2012-01-05.
[92] Motorola seeks to invalidate Apple phone patents (http://arstechnica.com/apple/news/2010/10/motorola-horns-in-on-apple-vs-htc-files-suit-against-apple.ars).
[93] Apple Files Lawsuit against Motorola to Defend Multi-Touch (http://www.patentlyapple.com/patently-apple/2010/10/apple-files-lawsuit-against-motorola-to-defend-multi-touch.html).
[94] Apple Patent Case Against Motorola to Get Trade Agency Review (http://www.businessweek.com/news/2010-11-23/apple-patent-case-against-motorola-to-get-trade-agency-review.html).
[95] Apple vs. Motorola: now 42 patents-in-suit (24 Apple and 18 Motorola patents) (http://fosspatents.blogspot.com/2010/12/apple-vs-motorola-now-42-patents-in.html).
[96] Microsoft sues Barnes & Noble over Android in Nook (http://www.geekwire.com/2011/breaking-microsoft-sues-barnes-noble-android).
[97] Nokia files second ITC complaint against Apple (http://press.nokia.com/2011/03/29/nokia-files-second-itc-complaint-against-apple/).
[98] FOSS Patents: Escalation: Nokia files new ITC complaint against Apple (plus a corresponding federal lawsuit) (http://fosspatents.blogspot.com/2011/03/escalation-nokia-files-new-itc.html).
[99] "Apple to Samsung: Stop stealing ideas" (http://www.cnn.com/2011/TECH/innovation/04/19/apple.samsung.lawsuit.wired/). *CNN.com*. . Retrieved 19 April 2011.
[100] Apple sues Samsung: a complete lawsuit analysis (http://thisismynext.com/2011/04/19/apple-sues-samsung-analysis/).
[101] Samsung Countersues Apple for Patent Infringement (http://www.pcmag.com/article2/0,2817,2383964,00.asp).
[102] Samsung sues Apple for infringing 10 patents: a closer look (http://thisismynext.com/2011/04/29/samsung-sues-apple-infringing-10-patents-closer/).
[103] Samsung Must Show Cell Phones to Apple (http://www.courthousenews.com/2011/05/19/36708.htm).
[104] Judge Orders Samsung to Give Apple Unreleased Tablets, Smartphones (http://www.pcmag.com/article2/0,2817,2385861,00.asp).
[105] Samsung's lawyers demand to see the iPhone 5 and iPad 3 (http://thisismynext.com/2011/05/28/samsung-apple-iphone-5-ipad-3/).
[106] Nokia Wins Apple Patent-License Deal Cash, Settles Lawsuits (http://www.bloomberg.com/news/2011-06-14/nokia-apple-payments-to-nokia-settle-all-litigation.html).
[107] Apple Settles With Nokia In Patent Lawsuit (http://www.huffingtonpost.com/2011/06/14/apple-nokia-patent-lawsuit-settlement_n_876499.html).
[108] Staff Reporter, "Apple alleges Samsung of 'slavishly copying' its technology; questions new Galaxy Tab 10.1" (http://www.ibtimes.com/articles/165405/20110619/apple-lawsuit-upgrade-samsung-galaxy-tab-10-1-infringement-intellectual-property-rights-ipad-2.htm), *International Business Times*, 19 June 2011.
[109] Shocker: Samsung not allowed to see iPhone 5 and iPad 3 (http://thisismynext.com/2011/06/22/shocker-samsung-allowed-iphone-5-ipad-3/).
[110] "Microsoft and General Dynamics Itronix Sign Patent Agreement: Agreement will cover General Dynamics Itronix devices running the Android platform" (http://www.microsoft.com/Presspass/press/2011/jun11/06-27ItronixPR.mspx). Microsoft.com. 2011-06-27. . Retrieved 2012-01-05.
[111] FOSS Patents: Android device makers sign up as Microsoft patent licensees -- an overview of the situation (http://fosspatents.blogspot.com/2011/06/android-device-makers-sign-up-as.html).

[112] "Microsoft and Velocity Micro, Inc., Sign Patent Agreement Covering Android-Based Devices: Agreement provides broad coverage of Microsoft's patent portfolio" (http://www.microsoft.com/Presspass/press/2011/jun11/06-29VelocityMicroPR.mspx). Microsoft.com. 2011-06-29. . Retrieved 2012-01-05.

[113] Dealtalk: Google bid "pi" for Nortel patents and lost (http://www.reuters.com/article/2011/07/02/us-dealtalk-nortel-google-idUSTRE76104L20110702).

[114] Nortel Announces the Winning Bidder of Its Patent Portfolio for a Purchase Price of US$4.5 Billion (http://www.marketwatch.com/story/nortel-announces-the-winning-bidder-of-its-patent-portfolio-for-a-purchase-price-of-us45-billion-2011-06-30?reflink=MW_news_stmp).

[115] Who Won The 6,000+ Nortel Patents? Apple, RIM, Microsoft — Everyone But Google (http://techcrunch.com/2011/07/01/apple-microsoft-rim-google-nortel-patents/).

[116] "Microsoft and Onkyo Corp. Sign Patent Agreement Covering Android-Based Tablets: Agreement provides broad coverage of Microsoft's patent portfolio" (http://www.microsoft.com/Presspass/press/2011/jun11/06-30OnkyoPR.mspx?rss_fdn=Custom). Microsoft.com. . Retrieved 2012-01-05.

[117] Apple files motion for preliminary injunction in the U.S. against four Samsung products: Infuse 4G, Galaxy S 4G, Droid Charge, Galaxy Tab 10.1 (http://fosspatents.blogspot.com/2011/07/apple-files-motion-for-preliminary.html).

[118] ITC ruling mixed in S3 Graphics v. Apple (http://news.cnet.com/8301-27076_3-20076219-248/itc-ruling-mixed-in-s3-graphics-v-apple/?tag=mncol;txt).

[119] "Microsoft and Wistron Sign Patent Agreement: Agreement will cover Wistron's Android tablets, smartphones and e-readers" (http://www.microsoft.com/Presspass/press/2011/jul11/07-05WistronPR.mspx). Microsoft.com. 2011-07-05. . Retrieved 2012-01-05.

[120] Microsoft patent division taking cash from at least 5 Android vendors (http://www.networkworld.com/news/2011/070511-microsoft-patent-android.html).

[121] DailyTech - VIA, WTI Sell Stakes in S3 Graphics to HTC (http://www.dailytech.com/VIA+WTI+Sell+Stakes+in+S3+Graphics+to+HTC/article22078.htm).

[122] HTC to Acquire Chip Designer S3 Graphics, Securing Patents in Apple Fight - Bloomberg (http://www.bloomberg.com/news/2011-07-06/htc-to-buy-s3-graphics-securing-patents-in-apple-fight-1-.html).

[123] HTC facing backlash for S3 acquisition (http://news.cnet.com/8301-1035_3-20077720-94/htc-facing-backlash-for-s3-acquisition/?tag=mncol;txt).

[124] Microsoft wants Samsung to pay smartphone license (http://www.reuters.com/article/2011/07/06/us-samsung-microsoft-idUSTRE7651DB20110706).

[125] HTC's Flyer Tablet, Droid Accused of Infringing Apple Patents - Bloomberg (http://www.bloomberg.com/news/2011-07-11/apple-files-new-trade-complaint-against-htc-over-devices.html?cmpid=yhoo).

[126] FOSS Patents: Apple files second ITC complaint against HTC: better luck next time? (http://fosspatents.blogspot.com/2011/07/apple-files-second-itc-complaint.html).

[127] Google Acquires Over 1,000 IBM Patents in July (http://www.seobythesea.com/2011/07/google-acquires-ibm-patents-in-july/).

[128] Google's New Patents from IBM (http://www.seobythesea.com/2011/07/googles-new-patents-from-ibm/).

[129] Bad News for Android: ITC Rules HTC Violated Two Apple Patents - John Paczkowski - News - AllThingsD (http://allthingsd.com/20110715/itc-rules-htc-violated-two-apple-patents/?refcat=news).

[130] Apple Lawsuit Puts Samsung Tablet Sales in Australia on Hold (http://www.bloomberg.com/news/2011-08-01/apple-seeks-to-block-samsung-from-selling-tablet-in-australia.html).

[131] Samsung's official comment on the Australian Galaxy Tab 10.1 situation is extremely weak (http://fosspatents.blogspot.com/2011/08/samsungs-official-comment-on-australian.html).

[132] Tsukayama, Hayley (10 August 2011). "Samsung Galaxy Tab 10.1 imports banned in most of Europe" (http://www.washingtonpost.com/blogs/faster-forward/post/samsung-galaxy-tab-101-imports-banned-in-most-of-europe/2011/08/10/gIQAmnVr6I_blog.html). *The Washington Post*. . Retrieved 11 August 2011.

[133] Preliminary injunction granted by German court: Apple blocks Samsung Galaxy Tab 10.1 in the entire European Union except for the Netherlands (http://fosspatents.blogspot.com/2011/08/preliminary-injunction-granted-by.html).

[134] "Microsoft files to ban imports of Motorola smartphones" (http://androidandme.com/2011/08/news/microsoft-files-to-ban-imports-of-motorola-smartphones). Android and Me. . Retrieved 2011-12-10.

[135] Womack, Brian (2011-08-22). "Motorola Value Found in 18 Patents Used Against Apple: Tech" (http://www.bloomberg.com/news/2011-08-22/motorola-s-value-for-google-found-in-18-patents-used-against-apple-tech.html). Bloomberg. . Retrieved 2011-12-10.

[136] "Samsung Galaxy tablet ban lifted in most of Europe" (http://uk.reuters.com/article/2011/08/16/tech-us-samsung-apple-ban-idUKTRE77F45420110816). *Reuters*. . Retrieved 16 August 2011.

[137] "Samsung Galaxy Tab ban is on hold" (http://www.bbc.co.uk/news/business-14548895). *BBC*. . Retrieved 16 August 2011.

[138] Google and IBM do it again: Google Acquires over 1,000 Patents from IBM in August (http://www.seobythesea.com/2011/09/google-ibm-patents-august/#more-6675).

[139] Microsoft Asks For An Import Ban On Motorola Smartphones (http://techcrunch.com/2011/08/23/microsoft-asks-for-an-import-ban-on-motorola-smartphones/).

[140] "Microsoft files to ban imports of Motorola smartphones" (http://androidandme.com/2011/08/news/microsoft-files-to-ban-imports-of-motorola-smartphones/). Android and Me. . Retrieved 2011-12-10.

[141] Euro ban for Samsung Galaxy phone (http://www.bbc.co.uk/news/technology-14652482).
[142] Samsung Puts Galaxy 10.1 Tablet on Hold as Apple Wins German Court Order (http://www.bloomberg.com/news/2011-09-04/apple-wins-german-injunction-on-new-galaxy-tab.html).
[143] FOSS Patents: Apple to ITC: Andy Rubin got inspiration for Android framework while working at Apple, hence infringes an Apple API patent (http://fosspatents.blogspot.com/2011/09/apple-to-itc-andy-rubin-got-inspiration.html).
[144] Milford, Phil (2011-09-08). "HTC Sues Apple Using Google Patents Bought Last Week as Battle Escalates" (http://www.bloomberg.com/news/2011-09-07/htc-sues-apple-alleging-infringement-of-four-u-s-patents.html). Bloomberg. . Retrieved 2011-12-10.
[145] Google Hands HTC Patents to Use Against Apple in Smartphone Wars (http://www.businessweek.com/news/2011-09-08/google-hands-htc-patents-to-use-against-apple-in-smartphone-wars.html).
[146] Patel, Nilay (2011-09-07). "HTC sues Apple for patent infringement... using patents purchased by Google" (http://thisismynext.com/2011/09/07/htc-sues-apple-patent-infringement-patents-purchased-google/). Thisismynext.com. . Retrieved 2012-01-05.
[147] Google gets its hands dirty - Apple 2.0 - Fortune Tech (http://tech.fortune.cnn.com/2011/09/08/google-gets-its-hands-dirty/).
[148] "Microsoft and Acer Sign Patent License Agreement: Agreement will cover Acer's Android tablets and smartphones" (http://www.microsoft.com/Presspass/press/2011/sep11/09-08AcerPR.mspx). Microsoft.com. 2011-09-08. . Retrieved 2012-01-05.
[149] "Microsoft and ViewSonic Sign Patent Agreement: Agreement will cover ViewSonic's Android Tablets and smartphones" (http://www.microsoft.com/Presspass/press/2011/sep11/09-08ViewSonicPR.mspx). Microsoft.com. 2011-09-08. . Retrieved 2012-01-05.
[150] "Microsoft ropes two more OEMs into Android patent deal" (http://www.zdnet.com/blog/hardware/microsoft-ropes-two-more-oems-into-android-patent-deal/14622). Zdnet.com. . Retrieved 2012-01-05.
[151] "Microsoft signs Android patent agreements with Acer, ViewSonic" (http://seattletimes.nwsource.com/html/microsoftpri0/2016144500_microsoft_signs_android_patent_agreements_with_ace.html). Seattletimes.nwsource.com. 2011-09-08. . Retrieved 2012-01-05.
[152] Matussek, Karin (2011-09-09). "Apple Wins German Ban on Samsung Tablet" (http://www.bloomberg.com/news/2011-09-09/apple-wins-ruling-for-german-samsung-galaxy-tablet-10-1-ban.html). Bloomberg.com. . Retrieved 2012-01-05.
[153] "Latest Samsung lawsuit targets Apple's iPhone, iPad in France" (http://www.appleinsider.com/articles/11/09/13/latest_samsung_lawsuit_targets_apples_iphone_ipad_in_france.html). AppleInsider. 2011-09-13. . Retrieved 2012-01-05.
[154] Meyer, David (2011-09-14). "Apple sues Samsung in the UK over Android | ZDNet UK" (http://www.zdnet.co.uk/blogs/communication-breakdown-10000030/apple-sues-samsung-in-the-uk-over-android-10024342/). Zdnet.co.uk. . Retrieved 2012-01-05.
[155] "USPTO Assignments on the Web" (http://assignments.uspto.gov/assignments/q?db=pat&reel=026894&frame=0001). Assignments.uspto.gov. . Retrieved 2012-01-05.
[156] "Samsung fires back at Apple in Australia with countersuit against iPhone, iPad" (http://www.appleinsider.com/articles/11/09/16/samsung_files_patent_case_against_apple_in_australia_over_iphone_ipad.html). Appleinsider.com. . Retrieved 2012-01-05.
[157] "Microsoft and Samsung Broaden Smartphone Partnership: Agreements mark new initiatives to promote Windows Phone and share intellectual property" (http://www.microsoft.com/Presspass/press/2011/sep11/09-28SamsungPR.mspx). Microsoft.com. 2011-09-28. . Retrieved 2012-01-05.
[158] Our Licensing Deal with Samsung: How IP Drives Innovation and Collaboration - Microsoft on the Issues - Site Home - TechNet Blogs (http://blogs.technet.com/b/microsoft_on_the_issues/archive/2011/09/28/our-licensing-deal-with-samsung-how-ip-drives-innovation-and-collaboration.aspx).
[159] Florian Mueller (2011-09-28). "FOSS Patents: Samsung takes Android patent license from Microsoft rather than wait for Motorola" (http://fosspatents.blogspot.com/2011/09/samsung-takes-android-patent-license.html). Fosspatents.blogspot.com. . Retrieved 2012-01-05.
[160] "Apple Wins Injunction Blocking Sale of Galaxy Tab 10.1 in Australia" (http://www.macrumors.com/2011/10/12/apple-wins-injunction-blocking-sale-of-galaxy-tab-10-1-in-australia/). Mac Rumors. 2011-10-12. . Retrieved 2012-01-05.
[161] "Microsoft and Quanta Computer Sign Patent Agreement Covering Android and Chrome-Based Devices - Redmond, Wash., Oct. 13, 2011 /PRNewswire/" (http://www.prnewswire.com/news-releases/microsoft-and-quanta-computer-sign-patent-agreement-covering-android-and-chrome-based-devices-131793888.html). Washington: Prnewswire.com. . Retrieved 2012-01-05.
[162] Microsoft Inks Another Android Patent Deal, This Time With Quanta | TechCrunch (http://techcrunch.com/2011/10/13/microsoft-inks-another-android-patent-deal-this-time-with-quanta/).
[163] Levine, Dan. "U.S. judge says Samsung tablets infringe Apple patents" (http://www.reuters.com/article/2011/10/13/us-apple-samsung-lawsuit-idUSTRE79C79C20111013?feedType=RSS&feedName=businessNews&utm_source=dlvr.it&utm_medium=twitter&dlvrit=56943). Reuters.com. . Retrieved 2012-01-05.
[164] Apple Inc.. "Distribute your App - iOS Developer Program - Apple Developer" (http://developer.apple.com/programs/ios/distribute.html). Developer.apple.com. . Retrieved 2012-01-05.
[165] "Apple's rivals battle for iOS scraps as app market sales grow to $2.2 billion" (http://www.appleinsider.com/articles/11/02/18/rim_nokia_and_googles_android_battle_for_apples_ios_scraps_as_app_market_sales_grow_to_2_2_billion.html). Appleinsider.com. 2011-02-18. . Retrieved 2012-01-05.
[166] "Google Android has double the number of free apps than Apple's App Store" (http://techcrunch.com/2010/07/05/distimo-june-2010/). Distimo. 15 July 2009. . Retrieved 15 July 2009.
[167] "McAfee: Android malware surges 76%, iPhone untouched" (http://www.electronista.com/articles/11/08/23/mcafee.shows.android.facing.huge.spike.in.malware/). Electronista.com. . Retrieved 2012-01-05.

[168] Dalrymple, Jim (2011-11-16). "Android sees a 472% increase in malware since July" (http://www.loopinsight.com/2011/11/16/android-sees-a-472-increase-in-malware-since-july/). Loopinsight.com. . Retrieved 2012-01-05.

[169] Mobile Malware Development Continues To Rise, Android Leads The Way (http://globalthreatcenter.com/?p=2492).

[170] "The Mother Of All Android Malware Has Arrived" (http://www.androidpolice.com/2011/03/01/the-mother-of-all-android-malware-has-arrived-stolen-apps-released-to-the-market-that-root-your-phone-steal-your-data-and-open-backdoor/). *Android Police*. March 6, 2011. .

[171] Perez, Sarah (2009-02-12). "Android Vulnerability So Dangerous, Owners Warned Not to Use Phone's Web Browser" (http://www.readwriteweb.com/archives/android_vulnerability_so_dangerous_shouldnt_use_web_browser.php). Readwriteweb.com. . Retrieved 2011-08-08.

[172] "Lookout, Retrevo warn of growing Android malware epidemic, note Apple's iOS is far safer" (http://www.appleinsider.com/articles/11/08/03/lookout_retrevio_warn_of_growing_android_malware_epidemic_note_apples_ios_is_far_safer.html). Appleinsider.com. 2011-08-03. . Retrieved 2012-01-05.

[173] "First SMS Trojan detected for smartphones running Android" (http://www.kaspersky.com/news?id=207576158). Kaspersky Lab. . Retrieved 2010-10-18.

[174] "Apple's iOS unaffected by malware as Android exploits surge 76%" (http://www.appleinsider.com/articles/11/08/24/apples_ios_unaffected_by_malware_as_android_exploits_surge_76.html). Appleinsider.com. 2011-08-24. . Retrieved 2012-01-05.

[175] "Security researcher finds code signing flaw that opens door for iOS malware" (http://www.appleinsider.com/articles/11/11/07/newly_found_code_signing_flaw_allows_for_ios_malware.html). Appleinsider.com. 2011-11-07. . Retrieved 2012-01-05.

[176] Apple's iOS more secure than Google's Android, says Symantec (http://www.appleinsider.com/articles/11/06/28/apples_ios_more_secure_than_googles_android_says_symantec.html)

[177] the understatement: Android Orphans: Visualizing a Sad History of Support (http://theunderstatement.com/post/11982112928/android-orphans-visualizing-a-sad-history-of-support)

[178] "Smart phones: how to stay clever in downturn" (http://www.deloitte.co.uk/TMTPredictions/telecommunications/Smartphones-clever-in-downturn.cfm). *Deloitte Telecommunications Predictions*. .

[179] "Android Phones Steal Market Share" (http://bmighty.informationweek.com/mobile/showArticle.jhtml?articleID=224201881). .

[180] "100 Million Club – H1 2010" (http://www.visionmobile.com/blog/2010/10/smart-feature-phones-the-unbalanced-equation-100-million-club-series/). .

[181] "Mobile Phone Development » Blog Archive » Gartner Q3 2010" (http://www.mobilephonedevelopment.com/archives/1149). Mobilephonedevelopment.com. 2010-11-10. . Retrieved 2011-09-07.

[182] Stan Schroeder (2011-02-10). "Gartner: Symbian Is Still the Number One Smartphone Platform" (http://mashable.com/2011/02/10/symbian-number-one-gartner-report/). Mashable.com. . Retrieved 2011-09-07.

[183] "Gartner Says Worldwide Mobile Device Sales to End Users Reached 1.6 Billion Units in 2010; Smartphone Sales Grew 72 Percent in 2010" (http://www.gartner.com/it/page.jsp?id=1543014). Gartner.com. 2011-02-11. . Retrieved 2011-09-07.

[184] "Olswang Predicts Mobile's Impact on TV: The Convergent Revolution" (http://www.tvgenius.net/blog/2011/03/17/mobile-tv-convergence/). Tvgenius.net. 2011-03-17. . Retrieved 2011-09-07.

[185] "Berg: Smartphone shipments grew 74% in 2010" (http://www.bgr.com/2011/03/10/berg-smartphone-shipments-grew-74-in-2010/). *Boy Genius Report*. March 10, 2011. .

[186] Generation App: 62% of Mobile Users 25-34 own Smartphones | Nielsen Wire (http://blog.nielsen.com/nielsenwire/?p=29786).

[187] Market Research | Consumer Market Research - NPD - As Smartphone Prices Fall, Retailers Are Leaving Money on the Table, According to The NPD Group (http://www.npdgroup.com/wps/portal/npd/us/news/pressreleases/pr_111114a)

[188] Apple, With 4 Percent of Handset Market, Captures 52 Percent of Profits (http://www.pcmag.com/article2/0,2817,2395951,00.asp#fbid=SLgcdJun7IU)

[189] "Smartphones killing point-and-shoots, now take almost 1/3 of photos" (http://gigaom.com/2011/12/22/smartphones-killing-point-and-shoots-now-take-almost-13-of-photos/). . Retrieved December 25, 2011.

[190] "Gartner Says Sales of Mobile Devices in Second Quarter of 2011 Grew 16.5 Percent Year-on-Year; Smartphone Sales Grew 74 Percent" (http://www.gartner.com/it/page.jsp?id=1764714). Gartner. 11 August 2011. . Retrieved 20 August 2011.

[191] Tarmo Virki (2011-01-31). "Google topples Symbian from smart phones top spot" (http://www.theglobeandmail.com/news/technology/mobile-technology/google-topples-symbian-from-smartphones-top-spot/article1888557/?cmpid=rss1). Theglobeandmail.com. . Retrieved 2011-09-07.

[192] "Android became clear smartphone leader in first quarter: Canalys" (http://www.reuters.com/article/2011/05/04/us-smartphones-research-idUSTRE7434VV20110504). Reuters.com. 2011-05-04. . Retrieved 2011-09-07.

[193] "Android Takes Over in UK" (http://www.bestsmartphone.com/2011/11/02/android-takes-over-in-uk/). bestsmartphone.com. 2011-11-02. .

[194] http://www.gartner.com/it/page.jsp?id=910112

[195] http://www.gartner.com/it/page.jsp?id=910112

[196] http://www.gartner.com/it/page.jsp?id=1306513

[197] http://www.gartner.com/it/page.jsp?id=1543014

[198] iPass : Mobile Workforce Report (http://mobile-workforce-project.ipass.com/).

[199] iPhone pushes past BlackBerry to top enterprise phone ranks (http://www.appleinsider.com/articles/11/11/16/iphone_pushes_past_blackberry_to_top_enterprise_phone_ranks.html)

[200] In the US Market, iPhone Outperforms Other Mobile Platforms in User Loyalty by a Wide Margin, Android is Second, Blackberry Fourth (http://www.zokem.com/2011/01/in-the-us-market-iphone-outperforms-other-mobile-platforms-in-user-loyalty-by-a-wide-margin-android-is-second-blackberry-fourth/)

[201] iPhone Whips Android, BlackBerry in User Loyalty: Zokem (http://www.eweek.com/c/a/Mobile-and-Wireless/iPhone-Whips-Android-Blackberry-in-User-Loyalty-Zokem-445681/)

[202] Strategy Analytics: Samsung Becomes World's Number One Smartphone Vendor in Q3 2011 - MarketWatch (http://www.marketwatch.com/story/strategy-analytics-samsung-becomes-worlds-number-one-smartphone-vendor-in-q3-2011-2011-10-27)

[203] http://www.gartner.com/it/page.jsp?id=1764714

[204] "Smartphone Sales Will Hit 420 Million In 2011, To Take 28 Percent Of The Total Phone Market" (http://techcrunch.com/2011/07/27/smartphone-sales-will-hit-420-million-in-2011-to-take-28-percent-of-the-total-phone-market/). TechCrunch. 27 July 2011. .

[205] Apple's iPhone accounted for 66% of Q2 smartphone profit among top vendors (http://www.bgr.com/2011/07/29/apples-iphone-accounted-for-66-of-q2-smartphone-profit-among-top-vendors/)

[206] "SA agrees: Apple now top smartphone vendor in the world with 140% growth" (http://www.bgr.com/2011/07/29/sa-agrees-apple-now-top-smartphone-vendor-in-the-world-with-240-growth/). BGR. 29 July 2011. .

[207] "Nokia reports Q2 results: Sells 88.5 million devices and declares net loss of 368 million euros" (http://mobilesyrup.com/2011/07/21/nokia-reports-q2-results-sells-88-5-million-devices-and-declares-net-loss-of-368-million-euros/). Mobilesyrup.com. 2011-07-21. . Retrieved 2011-09-07.

[208] "In U.S. Smartphone Market, Android is Top Operating System, Apple is Top Manufacturer" (http://blog.nielsen.com/nielsenwire/?p=28516). Nielsen. 28 July 2011. . Retrieved 5 September 2011.

[209] "Android Still Dominates Phones, But What About the Rest of Mobile?" (http://www.wired.com/gadgetlab/2011/07/android-ios-platform-share/all/1). Wired. 28 July 2011. . Retrieved 5 September 2011.

[210] iPhone 4 remains top-selling US smartphone despite growing iPhone 5 hype (http://www.appleinsider.com/articles/11/09/06/iphone_4_remains_top_selling_us_smartphone_despite_growing_iphone_5_hype.html)

[211] Previous-gen Apple iPad, iPhone 3GS often outsell new Android devices (http://www.appleinsider.com/articles/11/05/09/previous_gen_apple_ipad_iphone_3gs_often_outsell_new_android_devices.html)

[212] "Samsung Becomes Biggest Smartphone Vendor, as Android's Market Share Grows" (http://www.pcworld.com/article/243861/samsung_becomes_biggest_smartphone_vendor_as_androids_market_share_grows.html). PCWorld. 2011-11-15. . Retrieved 2011-12-10.

[213] Ever-Popular iPhone Named Top Smartphone. Again (http://www.wired.com/gadgetlab/2011/09/iphone-tops-survey-sixth-time/all/1).

[214] Wireless Consumer Smartphone Ratings (Volume 1) (http://www.jdpower.com/Electronics/ratings/wireless-consumer-smartphone-ratings-(volume-1)/)

[215] Wireless Consumer Smartphone Ratings (Volume 2) (http://www.jdpower.com/Electronics/ratings/wireless-consumer-smartphone-ratings-(volume-2)/).

[216] 2011 U.S. Wireless Handset Customer Satisfaction Studies—Vol. 2 (http://www.jdpower.com/news/pressRelease.aspx?ID=2011146)

[217] 2011 Wireless Smartphone and Traditional Mobile Phone Satisfaction Studies-Vol. 1 (http://www.jdpower.com/news/pressrelease.aspx?ID=2011030)

[218] 2010 U.S. Wireless Smartphone and Traditional Mobile Phone Customer Satisfaction Studies-Volume 1 (http://businesscenter.jdpower.com/news/pressrelease.aspx?ID=2010039)

[219] 2009 Wireless Consumer Smartphone and Traditional Mobile Phone Customer Satisfaction Studies (http://businesscenter.jdpower.com/news/pressrelease.aspx?ID=2009082)

[220] Menezes, Gary (2010-09-11). "Symbian OS, Now Fully Open Source" (http://www.watblog.com/2010/02/06/symbian-os-now-fully-open-source). Watblog.com. . Retrieved 2011-09-07.

[221] "Nokia N9 Smart Phone Review" (http://wontek.com/smart-phones/nokia/nokia-n9-smart-phone-review). Wontek.com. 2011-01-01. . Retrieved 2011-09-07.

[222] Lance Whitney, CNET. " Nearly 1 in 5 smartphone owners use check-in services (http://news.cnet.com/8301-1023_3-20062640-93.html?part=rss&subj=news&tag=2547-1_3-0-20)." May 13, 2011. Retrieved May 13, 2011.

[223] Second Screen change the way we watch TV (http://www.good.is/post/double-the-glow-will-second-screen-apps-change-the-way-we-watch-tv/)

[224] explosion of second screen apps (http://www.adweek.com/news/technology/second-screen-apps-explode-132237)

External links

- Freesmartphone.org (http://www.freesmartphone.org) – a collaboration platform for open-source smartphones
- Defining the Smartphone, part 1 (http://www.allaboutsymbian.com/features/item/Defining_the_Smartphone.php), part 2 (http://www.allaboutsymbian.com/features/item/Spy_versus_Spy_No_Smartphone_versus_Smartphone.php) by Steve Litchfield on July 16, 2010, various definitions are treated
- Mozilla Labs: Seabird –Community driven Mobile Phone Concept (http://mozillalabs.com/conceptseries/2010/09/23/seabird/)

ltg:Pruottelefons pfl:Smartphone

Series_90_(software_platform)

The **Series 90** (formerly *Hildon*) is a platform for mobile phones that uses Symbian OS. It was developed primarily by Nokia and is currently used only on the Nokia 7710 and Nokia 9500. Nokia discontinued Series 90 as a platform, but merged its technology into S60.[1] [2] Although no new Series 90 devices are expected, a form of the GUI lives on as Nokia's Hildon user interface in the Maemo shipped with Nokia Internet Tablets.

Compatibility

Series 90 is completely incompatible with Series 60 and UIQ, the most popular platforms for Symbian smartphones. However, some applications from Nokia Series 80 Communicator devices – such as the Nokia 9300 – will function under Series 90.

History

Essentially, Series 90 is a development of the Psion Eikon GUI as used on the Series 5, Series 5mx, Revo, netPad, Series 7 and netBook machines. It has been modified to be controlled entirely through the touchscreen, supplemented by the 77x0's seven hardware buttons and rocker-dial. The Nokia 7700, which was never released, had only an on-screen keyboard for text input; the 7710 added basic handwriting recognition. Applications now have only a single menu tree, accessible from the title bar at the top of the screen; there is no Exit or Quit option in most apps; and a button bar has been added to the right hand side of the screen, resembling that of the Nokia Communicator smartphones.

References

[1] Forum Nokia Wiki – Series 90 as a platform is discontinued and has been merged into S60 (http://wiki.forum.nokia.com/index.php/Series_90)

[2] IntoMobile: *The Series 90 future* (http://www.intomobile.com/2006/04/13/the-series-90-future.html)

External links

- Forum Nokia Wiki - Series 90 (http://wiki.forum.nokia.com/index.php/Series_90)

OMAP

OMAP developed by **Texas Instruments** is a category of proprietary system on chips (SoCs) for portable and mobile multimedia applications. OMAP devices generally include a general-purpose ARM architecture processor core plus one or more specialized co-processors. Earlier OMAP variants commonly featured a variant of the Texas Instruments TMS320 series digital signal processor.

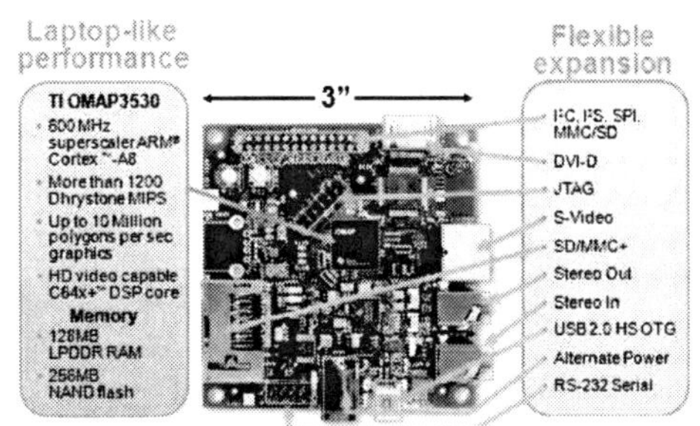

TI OMAP3530 on BeagleBoard described

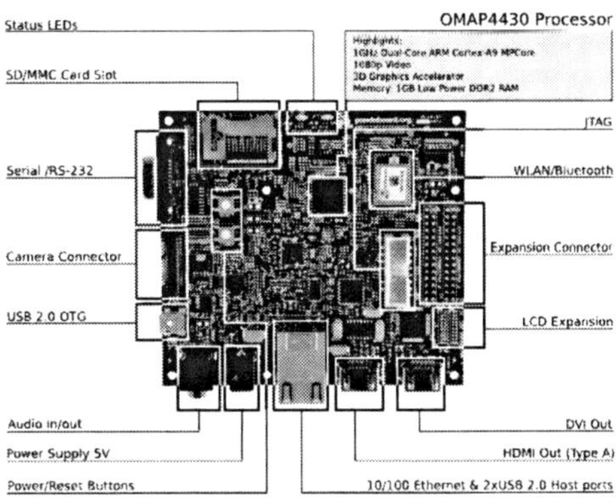

TI OMAP4430 on PandaBoard described

OMAP family

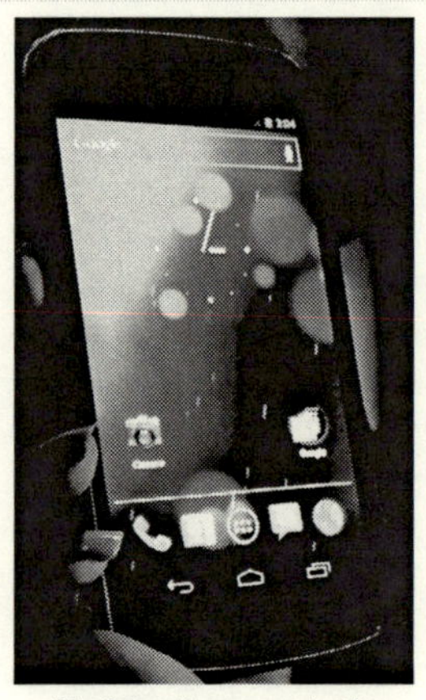

The Galaxy Nexus, example of a smartphone with an OMAP 4460 SoC

The OMAP family consists of three product groups classified by performance and intended application:

- High-performance applications processors
- Basic multimedia applications processors
- Integrated modem and applications processors

Further, two main distribution channels exist, and not all parts are available in both channels. The genesis of the OMAP product line is from partnership with cell phone vendors, and the main distribution channel involves sales directly to such *wireless handset* vendors. Parts developed to suit evolving cell phone requirements are flexible and powerful enough to support sales through less specialized *catalog* channels; some OMAP 1 parts, and many OMAP 3 parts, have catalog versions with different sales and support models. Parts that are obsolete from the perspective of handset vendors may still be needed to support products developed using catalog parts and distributor-based inventory management.

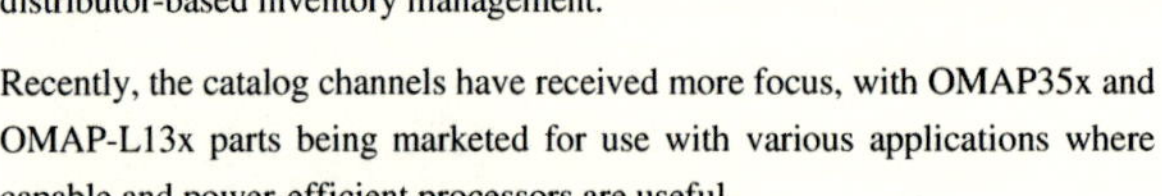

Recently, the catalog channels have received more focus, with OMAP35x and OMAP-L13x parts being marketed for use with various applications where capable and power-efficient processors are useful.

High-performance applications processors

These are parts originally intended for use as application processors in smartphones, with processors powerful enough to run significant operating systems (such as Linux, Android or Symbian), support connectivity to personal computers, and support various audio and video applications.

OMAP 1

The OMAP 1 family started with a TI-enhanced ARM core, and then changed to a standard ARM926 core. It included many variants, most easily distinguished according to manufacturing technology (130 nm except for the OMAP171x series), CPU, peripheral set, and distribution channel (direct to large handset vendors, or through catalog-based distributors). In March 2009, the OMAP1710 family chips are still available to handset vendors.

Products using OMAP 1 processors include hundreds of cell phone models, and the Nokia 770 Internet tablets.

- OMAP171x - 220 MHz ARM926EJ-S + C55x DSP, low-voltage 90 nm technology
- OMAP162x - 204 MHz ARM926EJ-S + C55x DSP + 2 MB internal SRAM, 130 nm technology
- OMAP5912 - catalog availability version of OMAP1621 (or OMAP1611b in older versions)
- OMAP161x - 204 MHz ARM926EJ-S + C55x DSP, 130 nm technology
- OMAP1510 - 168 MHz ARM925T (TI-enhanced) + C55x DSP
- OMAP5910 - catalog availability version of OMAP 1510

OMAP 2

These parts were only marketed to handset vendors. Products using these include both Internet tablets and mobile phones:

- OMAP2431 - 330 MHz ARM1136 + 220 MHz C64x DSP
- OMAP2430 - 330 MHz ARM1136 + 220 MHz C64x DSP + PowerVR MBX lite GPU
- OMAP2420 - 330 MHz ARM1136 + 220 MHz C55x DSP + PowerVR MBX GPU

OMAP 3

The 3rd generation OMAP, The OMAP 3[1] is broken into 3 distinct groups: the OMAP34x, the OMAP35x, and the OMAP36x. OMAP34x and OMAP36x are distributed directly to large handset (such as cell phone) manufacturers. OMAP35x is a variant of OMAP34x intended for catalog distribution channels. The OMAP36x is a 45 nm version of the 65 nm OMAP34x with higher clock speed.[2]

The video technology in the higher end OMAP 3 parts is derived in part from the DaVinci product line, which first packaged higher end C64x+ DSPs and image processing controllers with ARM9 processors last seen in the older OMAP 1 generation or ARM Cortex-A8[3] .

Not highlighted in the list below is that each OMAP 3 SoC has an "Image, Video, Audio" (IVA2) accelerator. These units do not all have the same capabilities. Most devices support 12 megapixel camera images, though some support 5 or 3 megapixels. Some support HD imaging.

Model number	Semiconductor technology	CPU instruction set	CPU	GPU	Utilizing devices
OMAP3410	65 nm	ARMv7	600 MHz ARM Cortex-A8	PowerVR SGX530	Motorola Charm, Motorola Flipout, Motorola Flipside
OMAP3420	65 nm	ARMv7	600 MHz ARM Cortex-A8	PowerVR SGX530	
OMAP3430	65 nm	ARMv7	600 MHz ARM Cortex-A8	PowerVR SGX530	Motorola Droid/Milestone, Palm Pre, Samsung i8910, Nokia N900
OMAP3440	65 nm	ARMv7	800 MHz ARM Cortex-A8	PowerVR SGX530	Motorola XT720, Archos 5 (Gen 7), Samsung SHW-M100S Galaxy A, Motorola Titanium XT800
OMAP3503	65 nm	ARMv7	600 MHz ARM Cortex-A8	N/A	Gumstix Overo Earth
OMAP3515	65 nm	ARMv7	600 MHz ARM Cortex-A8	PowerVR SGX530	
OMAP3525	65 nm	ARMv7	600 MHz ARM Cortex-A8	N/A	
OMAP3530	65 nm	ARMv7	720 MHz ARM Cortex-A8	PowerVR SGX530	Tianyeit Cip312 Computer In Package [4],phyCARD-L System on Module [5], BeagleBoard, Gumstix, IGEPv2, Alico's Kinetic 3500,[6] OSWALD, Overo Water, Pandora, Touch Book, Embest DevKit8000 [7], OpenSourceMID [8].
OMAP3611[9]	45 nm	ARMv7	800 MHz ARM Cortex-A8	PowerVR SGX530	Cybook Odyssey
OMAP3621(OMAP3622)	45 nm	ARMv7	3621: 800 MHz, 3622: 1 GHz; ARM Cortex-A8	PowerVR SGX530	Nook Color, Nook Simple Touch, IdeaPad A1

OMAP3630	45 nm	ARMv7	600 MHz~1.2 GHz ARM Cortex-A8	PowerVR SGX530	3630-800MHz(underclocked): Motorola Bravo, Motorola Defy;[10] 3630-1000MHz(Default Speed): Nokia N9, Nokia N950, Sony Ericsson Vivaz, Motorola Milestone 2,Motorola Cliq 2, Motorola Defy+, Palm Pre 2, Droid X, Droid 2, Archos 101, Archos 70, Archos 43, Archos 32, Archos 28, LG Optimus Black, LG Optimus bright L-07C, Samsung I9003 Galaxy S(C)L, SAMSUNG GALAXY 7 TAB (P1010) Wifi Only, LG-LU3000 Optimus Mach, Panasonic P-07C, Panasonic Sweety 003P 3630-1200MHz(overclocked): Motorola Droid 2 Global

OMAP 4

The 4th generation OMAPs, OMAP 4430, 4460 (formerly named 4440),[11] and 4470 all use dual-core ARM Cortex-A9s. The 4470 additionally contains two Cortex-M3s running at 266 MHz to offload the A9s in less computionally intensive tasks to increase power efficiency.[12] [13] [14] 4430 and 4460 use a PowerVR SGX540 integrated 3D graphics accelerator which runs at a clock frequency of 304 and 384 MHz respectively compared to prior versions of SGX540 typically at 200 MHz making them theoretically much faster.[15] 4470 has a PowerVR SGX544 GPU that supports DirectX 9 which enables it for use in Windows 8 as well as a dedicated 2D graphics core for increased power efficiency. All OMAP 4 comes with an IVA3 multimedia hardware accelerator with a programmable DSP that enables 1080p Full HD and multi-standard video encode/decode.[16] [17] [18] [19] [20] OMAP 4 uses ARM-Cortex A9s with ARMs SIMD engine (Media Processing Engine, aka NEON) which may have a significant performance advantage in some cases over Nvidia Tegra 2s Cortex-A9s with non-vector floating point units.[21] It also uses a dual-channel LPDDR2 memory controller compared to Nvidia Tegra 2s single-channel memory controller.

Model number	Semiconductor technology	CPU instruction set	CPU	GPU	Memory technology	Availability	Utilizing devices
OMAP4430	45 nm	ARMv7	1-1.2 GHz dual-core ARM Cortex-A9	PowerVR SGX540 @ 304 MHz	Dual-channel LPDDR2 memory controller	Q1 2011	phyCORE-OMAP4430 System on Module [22], CIP410 based on OMAP4430 Computer In Package [4],PandaBoard, RIM BlackBerry Playbook[23] , LG Optimus 3D P920, Motorola Droid/Milestone 3, Motorola Droid Bionic, Motorola ATRIX 2, Motorola Droid RAZR[24] , Toshiba AT200 Excite, Fujitsu ARROWS Tab LTE F-01D, Fujitsu ARROWS Z ISW11F, Fujitsu ARROWS X LTE F-05D, Fujitsu REGZA Phone T-01D, Kindle Fire, Panasonic LUMIX Phone P-02D, Panasonic LUMIX Phone 101P, Archos 101(Gen 9), Archos 80(Gen 9), Barnes and Noble Nook tablet, SHARP AQUOS PHONE SH-01D, SHARP AQUOS PHONE 102SH, Raymarine e7 Multifunction Display [25], Samsung Galaxy S II I9100G

OMAP4460	45 nm	ARMv7	1.2-1.5 GHz dual-core ARM Cortex-A9	PowerVR SGX540 @ 384 MHz	Dual-channel LPDDR2 memory controller	Q4 2011	SHARP AQUOS PHONE 104SH, Galaxy Nexus, Archos 101 Turbo(Gen 9) [26], Archos 80 Turbo(Gen 9), Variscite VAR-SOM-OM44 System on Module [27], Pandaboard ES [28]
OMAP4470	45 nm	ARMv7	1.5-1.8 GHz dual-core ARM Cortex-A9	PowerVR SGX544 @ 384 MHz + dedicated 2D graphics core	Dual-channel LPDDR2 memory controller, 466 MHz	Q2 2012	

OMAP 5

The 5th generation OMAP, OMAP 5 SoC uses a dual-core ARM Cortex-A15 CPU with two additional Cortex-M4 cores to offload the A15s in less computionally intensive tasks to increase power efficiency, two PowerVR SGX544MP graphics cores and a dedicated TI 2D BitBlt graphics accelerator, a multi-pipe display sub-system and a signal processor.[29] They respectively support 24 and 20 megapixel cameras for front and rear 3D HD video recording. The chip also supports up to 8 GB of dual channel DDR3 memory, output to four HD 3D displays and 3D HDMI 1.4 video output. OMAP 5 also includes 3 USB 2.0 ports and a SATA 2.0 controller.

Model number	Semiconductor technology	CPU instruction set	CPU	GPU	Memory technology	Availability
OMAP5430	28 nm	ARMv7	2 GHz dual-core ARM Cortex-A15	Dual-core PowerVR SGX544MP + dedicated 2D graphics accelerator	Dual-channel package on package LPDDR2	Q3 2012
OMAP5432	28 nm	ARMv7	2 GHz dual-core ARM Cortex-A15	Dual-core PowerVR SGX544MP + dedicated 2D graphics accelerator	Dual-channel DDR3 controller	Q3 2012

Basic multimedia applications processors

These are marketed only to handset manufacturers. They are intended to be highly integrated, low cost chips for consumer products. The OMAP-DM series are intended to be used as digital media coprocessors for mobile devices with high megapixel digital still and video cameras.

- OMAP331 - ARM9
- OMAP310 - ARM9
- OMAP-DM270 - ARM7 + C54x DSP
- OMAP-DM299 - ARM7 + ISP + stacked mDDR SDRAM
- OMAP-DM500 - ARM7 + ISP + stacked mDDR SDRAM
- OMAP-DM510 - ARM926 + ISP + 128 MB stacked mDDR SDRAM
- OMAP-DM515 - ARM926 + ISP + 256 MB stacked mDDR SDRAM
- OMAP-DM525 - ARM926 + ISP + 256 MB stacked mDDR SDRAM

Integrated modem and applications processors

These are marketed only to handset manufacturers. Many of the newer versions are highly integrated for use in very low cost cell phones.

An OMAP 850 in a HTC Wizard

- OMAPV1035 - single-chip EDGE (was discontinued in 2009 as TI announced baseband chipset market withdrawal).
- OMAPV1030 - EDGE digital baseband
- OMAP850 - 200 MHz ARM926EJ-S + GSM/GPRS digital baseband + stacked EDGE co-processor
- OMAP750 - 200 MHz ARM926EJ-S + GSM/GPRS digital baseband + DDR Memory support
- OMAP733 - 200 MHz ARM926EJ-S + GSM/GPRS digital baseband + stacked SDRAM
- OMAP730 - 200 MHz ARM926EJ-S + GSM/GPRS digital baseband + SDRAM Memory support
- OMAP710 - 133 MHz ARM925 + GSM/GPRS digital baseband

OMAP L-1x

The OMAP L-1x parts are marketed only through catalog channels, and have a different technological heritage than the other OMAP parts. Rather than deriving directly from cell phone product lines, they grew from the video-oriented DaVinci product line by removing the video-specific features while using upgraded DaVinci peripherals. A notable feature is use of a *floating point* DSP, instead of the more customary fixed point one.

The Hawkboard uses the OMAP-L138

- OMAP-L137 - 300 MHz ARM926EJ-S + C674x floating point DSP
- OMAP-L138 - 300 MHz ARM926EJ-S + C674x floating point DSP

Products using OMAP processors

Many mobile phones use OMAP SoCs, including the Nokia N90, N91, N92, N95, N82, E61, E62, E63 and E90 mobile phones, as well as the N800, N810 and N900 Internet tablets, Motorola Droid, Droid X, and Droid 2. The Palm Pre, Pandora, Touch Book also use an OMAP SoC (the OMAP3430). Others to use an OMAP SoC include Sony Ericsson's Satio and Vivaz, the Samsung Omnia HD, Sony Ericsson Idou, the Nook Color, Kindle Fire, and some Archos tablets (such as Archos 80 gen 9 and Archos 101 gen 9).

OMAP SoCs are also used as the basis for a number of hobbyist and prototyping boards, such as the Beagle Board and Panda Board.

Similar platforms

- Snapdragon by Qualcomm
- Tegra by Nvidia
- Exynos by Samsung
- Ax by Apple
- NovaThor by ST-Ericsson

See also

- OpenMAX IL (Open Media Acceleration Integration Layer) - a royalty-free cross-platform media abstraction API from the Khronos Group
- Distributed Codec Engine (libcde) is a Texas Instruments API for the video codec engine in OMAP based embedded systems

References

[1] OMAP34xx series in TI Web site (http://focus.ti.com/general/docs/wtbu/wtbuproductcontent.tsp?templateId=6123&navigationId=11989&contentId=4682)
[2] OMAP36x (http://linuxdevices.com/news/NS8848023892.html)
[3] DaVinci Digital Video Processor - TMS320DM37x SOC - DM3730 - TI.com (http://www.ti.com/product/dm3730)
[4] http://www.tianyeit.com
[5] http://www.phytec.com/products/som/Cortex-A8/phyCARD-L-CortexA8-OMAP3525.html
[6] http://www.alicosystems.com/Alico%20FSDK%203500%200311A.pdf
[7] http://www.armkits.com/Product/devkit8000.asp
[8] http://www.opensourcemid.org
[9] The 3611 is the same die as 3621 but with restricted access to the DSP part
[10] http://focus.ti.com/pdfs/wtbu/OMAP36xx_ES1.x_PUBLIC_TRM_vN.zip
[11] http://www.linuxfordevices.com/c/a/News/Variscite-VARSOMOM44/Computer module taps 1.5GHz, dual-core OMAP4460 SoC
[12] http://focus.ti.com/pdfs/wtbu/OMAP4430_ES2.x_Public_TRM_vK.zip
[13] "OMAP4460 Public TRM vE (pdf)" (http://focus.ti.com/pdfs/wtbu/OMAP4460_ES1.0_PUBLIC_TRM_vE.zip)
[14] Texas Instruments announces multi-core, 1.8GHz OMAP4470 ARM processor for Windows 8 - Engadget (http://www.engadget.com/2011/06/02/texas-instruments-announces-multi-core-1-8ghz-omap4470-arm-proc/)
[15] AnandTech - TI Announces OMAP4470 and Specs: PowerVR SGX544, 1.8 GHz Dual Core Cortex-A9 (http://www.anandtech.com/show/4413/ti-announces-omap-4470-and-specs-powervr-sgx544-18-ghz-dual-core-cortexa9)
[16] OMAP44xx series in TI Web site (http://focus.ti.com/general/docs/wtbu/wtbuproductcontent.tsp?templateId=6123&navigationId=12843&contentId=53243)
[17] http://www.linuxfordevices.com/c/a/News/TI-OMAP4430-and-OMAP4440/ TI speeds up its OMAP 4 for 3D video
[18] http://www.engadget.com/2010/02/02/tis-omap4-prototype-drives-three-independent-displays-without-b/ TI's OMAP 4 prototype drives three independent displays without breaking a sweat
[19] http://www.engadget.com/2009/02/17/tis-omap-4-bringing-1080p-support-to-smartphones-and-mids/TI's OMAP 4 bringing 1080p support to smartphones and MIDs
[20] http://www.engadget.com/2010/02/15/texas-instruments-introduces-arm-based-omap-4-soc-blaze-develop/ Texas Instruments introduces ARM-based OMAP 4 SOC, Blaze development platform
[21] AnandTech - NVIDIA's Tegra 2 Take Two: More Architectural Details and Design Wins (http://www.anandtech.com/show/4098/nvidias-tegra-2-take-two-more-architectural-details-and-design-wins/2)
[22] http://www.phytec.com/products/som/Cortex-A8-A9/phyCORE-OMAP4430.html
[23] Blackberry confirms PlayBook specs and launch date - Telegraph (http://www.telegraph.co.uk/technology/blackberry/8400612/Blackberry-confirms-PlayBook-specs-and-launch-date.html)
[24] MOTODEV > Products > DROID RAZR™ by Motorola, XT912 (http://developer.motorola.com/products/droid-razr-xt912/)
[25] http://www.raymarine.co.uk/view/?id=1347
[26] http://www.archos.com/products/gen9/archos_101g9/specs.html?country=us&lang=en
[27] http://www.variscite.com/products/item/76-var-som-om44-ti-omap4460
[28] http://pandaboard.org/node/300/#PandaES
[29] "Not Just a Faster Horse: TI's OMAP 5 Platform Transforms the Concept of 'Mobile'" (http://newscenter.ti.com/Blogs/newsroom/archive/2011/02/07/not-just-a-faster-horse-ti-s-omap-5-platform-transforms-the-concept-of-mobile-615064.aspx). Texas Instruments. 11-02-07. . Retrieved 2011-02-09. "The OMAP 5 processor leverages two ARM Cortex-A15 MPCores [...] [It] also includes two ARM Cortex-M4 processors [...]"

External links

- OMAP Application Processors (http://www.ti.com/omap)
- OMAPWorld (http://www.omapworld.org/)
- OMAPpedia (http://www.omappedia.org/wiki/Main_Page)
- Linux OMAP Mailing List Archive (http://www.spinics.net/lists/linux-omap/)
- OMAP3 Boards (http://wiki.omap.com/index.php/OMAP3_Boards)
- OMAP4 Boards (http://wiki.omap.com/index.php/OMAP4_Boards)

Synchronous_dynamic_random-access_memory

Synchronous dynamic random access memory (**SDRAM**) is dynamic random access memory (DRAM) that is synchronized with the system bus. Classic DRAM has an asynchronous interface, which means that it responds as quickly as possible to changes in control inputs. SDRAM has a synchronous interface, meaning that it waits for a clock signal before responding to control inputs and is therefore synchronized with the computer's system bus. The clock is used to drive an internal finite state machine that pipelines incoming instructions. This allows the chip to have a more complex pattern of operation than an asynchronous DRAM, enabling higher speeds.

Pipelining means that the chip can accept a new instruction before it has finished processing the previous one. In a pipelined write, the write command can be immediately followed by another instruction without waiting for the data to be written to the memory array. In a pipelined read, the requested data appears after a fixed number of clock pulses after the read instruction, cycles during which additional instructions can be sent. (This delay is called the *latency* and is an important parameter to consider when purchasing SDRAM for a computer.)

SDRAM is widely used in computers; from the original SDRAM, further generations of DDR (or *DDR1*) and then DDR2 and DDR3 have entered the mass market, with DDR4 currently being designed and anticipated to be available in 2015.

SDRAM history

Although the concept of synchronous DRAM has been known since at least the 1970s and was used with early Intel processors, it was only in 1993 that SDRAM began its path to universal acceptance in the electronics industry. In 1993, Samsung introduced its KM48SL2000 synchronous DRAM, and by 2000, SDRAM had replaced virtually all other types of DRAM in modern computers, because of its greater performance.

Eight SDRAM ICs on a PC100 DIMM package.

SDRAM latency is not inherently lower (faster) than asynchronous DRAM. Indeed, early SDRAM was somewhat slower than contemporaneous burst EDO DRAM due to the additional logic. The benefits of SDRAM's internal buffering come from its ability to interleave operations to multiple banks of memory, thereby increasing effective bandwidth.

Today, virtually all SDRAM is manufactured in compliance with standards established by JEDEC, an electronics industry association that adopts open standards to facilitate interoperability of electronic components. JEDEC formally adopted its first SDRAM standard in 1993 and subsequently adopted other SDRAM standards, including those for DDR, DDR2 and DDR3 SDRAM.

SDRAM is also available in registered varieties, for systems that require greater scalability such as servers and workstations.

As of 2007, 168-pin SDRAM DIMMs are not used in new PC systems, and 184-pin DDR memory has been mostly superseded. DDR2 SDRAM is the most common type used with new PCs, and DDR3 motherboards and memory are widely available, and less expensive than still-popular DDR2 products.

Today, the world's largest manufacturers of SDRAM include: Samsung Electronics, Panasonic, Micron Technology, and Hynix.

SDRAM timing

There are several limits on DRAM performance. Most noted is the read cycle time, the time between successive read operations to an open row. This time decreased from 10 ns for 100 MHz SDRAM to 5 ns for DDR-400, but has remained relatively unchanged through DDR2-800 and DDR3-1600 generations. However, by operating the interface circuitry at increasingly higher multiples of the fundamental read rate, the achievable bandwidth has increased rapidly.

Another limit is the CAS latency, the time between supplying a column address and receiving the corresponding data. Again, this has remained relatively constant at 10–15 ns through the last few generations of DDR SDRAM.

In operation, CAS latency is a specific number of clock cycles programmed into the SDRAM's mode register and expected by the DRAM controller. Any value may be programmed, but the SDRAM will not operate correctly if it is too low. At higher clock rates, the useful CAS latency in clock cycles naturally increases. 10–15 ns is 2–3 cycles (CL2–3) of the 200 MHz clock of DDR-400 SDRAM, CL4-6 for DDR2-800, and CL8-12 for DDR3-1600. Slower clock cycles will naturally allow lower numbers of CAS latency cycles.

SDRAM modules have their own timing specifications, which may be slower than those of the chips on the module. When 100 MHz SDRAM chips first appeared, some manufacturers sold "100 MHz" modules that could not reliably operate at that clock rate. In response, Intel published the PC100 standard, which outlines requirements and guidelines for producing a memory module that can operate reliably at 100 MHz. This standard was widely influential, and the term "PC100" quickly became a common identifier for 100 MHz SDRAM modules, and modules are now commonly designated with "PC"-prefixed numbers (PC66, PC100 or PC133 - although the actual meaning of the numbers has changed).

SDR SDRAM

Originally simply known as *SDRAM*, **single data rate** SDRAM can accept one command and transfer one word of data per clock cycle. Typical clock frequencies are 100 and 133 MHz. Chips are made with a variety of data bus sizes (most commonly 4, 8 or 16 bits), but chips are generally assembled into 168-pin DIMMs that read or write 64 (non-ECC) or 72 (ECC) bits at a time.

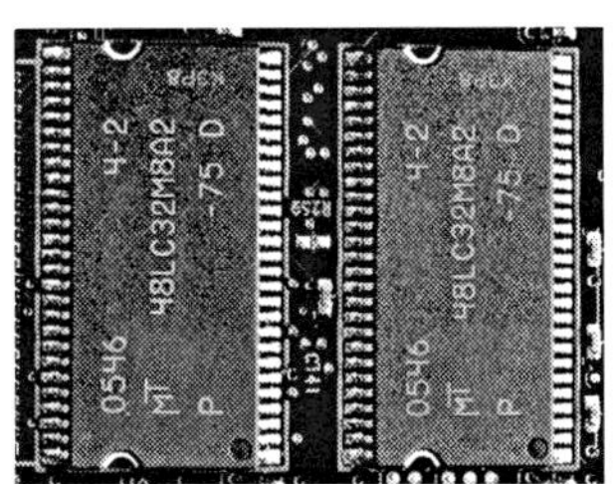

64 MB sound memory of Sound Blaster X-Fi Fatal1ty Pro uses two Micron 48LC32M8A2-75 C SDRAM chips working at 133 MHz (7.5 ns) 8-bit wide [1]

Use of the data bus is intricate and thus requires a complex DRAM controller circuit. This is because data written to the DRAM must be presented in the same cycle as the write command, but reads produce output 2 or 3 cycles after the read command. The DRAM controller must ensure that the data bus is never required for a read and a write at the same time.

Typical SDR SDRAM clock rates are 66, 100, and 133 MHz (periods of 15, 10, and 7.5 ns). Clock rates up to 150 MHz were available for performance enthusiasts.

SDRAM control signals

All commands are timed relative to the rising edge of a clock signal. In addition to the clock, there are 6 control signals, mostly active low, which are sampled on the rising edge of the clock:

- **CKE** Clock Enable. When this signal is low, the chip behaves as if the clock has stopped. No commands are interpreted and command latency times do not elapse. The state of other control lines is not relevant. The effect of this signal is actually delayed by one clock cycle. That is, the current clock cycle proceeds as usual, but the following clock cycle is ignored, except for testing the CKE input again. Normal operations resume on the rising edge of the clock after the one where CKE is sampled high.
  Put another way, all other chip operations are timed relative to the rising edge of a masked clock. The masked clock is the logical AND of the input clock and the state of the CKE signal during the previous rising edge of the input clock.
- **/CS** Chip Select. When this signal is high, the chip ignores all other inputs (except for CKE), and acts as if a NOP command is received.
- **DQM** Data Mask. (The letter Q appears because, following digital logic conventions, the data lines are known as "DQ" lines.) When high, these signals suppress data I/O. When accompanying write data, the data is not actually written to the DRAM. When asserted high two cycles before a read cycle, the read data is not output from the chip. There is one DQM line per 8 bits on a x16 memory chip or DIMM.
- **/RAS** Row Address Strobe. Despite the name, this is *not* a strobe, but rather simply a command bit. Along with /CAS and /WE, this selects one of 8 commands.
- **/CAS** Column Address Strobe. Despite the name, this is *not* a strobe, but rather simply a command bit. Along with /RAS and /WE, this selects one of 8 commands.
- **/WE** Write enable. Along with /RAS and /CAS, this selects one of 8 commands. This generally distinguishes read-like commands from write-like commands.

SDRAM devices are internally divided into 2 or 4 independent internal data banks. One or two bank address inputs (BA0 and BA1) select which bank a command is directed toward.

Many commands also use an address presented on the address input pins. Some commands, which either do not use an address, or present a column address, also use A10 to select variants.

The commands understood are as follows:

/CS	/RAS	/CAS	/WE	BA*n*	A10	A*n*	Command
H	x	x	x	x	x	x	Command inhibit (No operation)
L	H	H	H	x	x	x	No operation
L	H	H	L	x	x	x	Burst Terminate: stop a burst read or burst write in progress.
L	H	L	H	bank	L	column	Read: Read a burst of data from the currently active row.
L	H	L	H	bank	H	column	Read with auto precharge: As above, and precharge (close row) when done.
L	H	L	L	bank	L	column	Write: Write a burst of data to the currently active row.
L	H	L	L	bank	H	column	Write with auto precharge: As above, and precharge (close row) when done.
L	L	H	H	bank	row		Active (activate): open a row for Read and Write commands.
L	L	H	L	bank	L	x	Precharge: Deactivate current row of selected bank.
L	L	H	L	x	H	x	Precharge all: Deactivate current row of all banks.
L	L	L	H	x	x	x	Auto refresh: Refresh one row of each bank, using an internal counter. All banks must be precharged.

L	L	L	L	0 0	mode	Load mode register: A0 through A9 are loaded to configure the DRAM chip. The most significant settings are CAS latency (2 or 3 cycles) and burst length (1, 2, 4 or 8 cycles)

The various DDRx SDRAM standards use essentially the same commands, with minor additions. Additional mode registers are distinguished using the bank address bits, and a third bank address bit is added.

SDRAM operation

A 512 MB SDRAM DIMM (which contains 512 MiB = $512^2 \times 1024^2$ bytes = 536,870,912 bytes exactly) might be made of 8 or 9 SDRAM chips, each containing 512 Mbit of storage, and each one contributing 8 bits to the DIMM's 64- or 72-bit width. A typical 512 Mbit SDRAM chip internally contains 4 independent 16 Mbyte memory banks. Each bank is an array of 8,192 rows of 16,384 bits each. A bank is either idle, active, or changing from one to the other.

The Active command activates an idle bank. It presents a 2-bit bank address (BA0 BA1) and a 13-bit row address (A0 A12), and causes a read of that row into the bank's array of all 16,384 column sense amplifiers. This is also known as "opening" the row. This operation has the side effect of refreshing the dynamic (capacitive) memory storage cells of that row.

Once the row has been activated or "opened", Read and Write commands are possible to that row. Activation requires a minimum amount of time, called the row-to-column delay, or t_{RCD} before reads or writes to it may occur. This time, rounded up to the next multiple of the clock period, specifies the minimum number of wait cycles between an Active command, and a Read or Write command. During these wait cycles, additional commands may be sent to other banks; because each bank operates completely independently.

Both Read and Write commands require a column address. Because each chip accesses 8 bits of data at a time, there are 2048 possible column addresses thus requiring only 11 address lines (A0 A9, A11).

When a Read command is issued, the SDRAM will produce the corresponding output data on the DQ lines in time for the rising edge of the clock 2 or 3 clock cycles later (depending on the configured CAS latency). Subsequent words of the burst will be produced in time for subsequent rising clock edges.

A Write command is accompanied by the data to be written driven on to the DQ lines during the same rising clock edge. It is the duty of the memory controller to ensure that the SDRAM is not driving read data on to the DQ lines at the same time that it needs to drive write data on to those lines. This can be done by waiting until a read burst has finished, by terminating a read burst, or by using the DQM control line.

When the memory controller needs to access a different row, it must first return that bank's sense amplifiers to an idle state, ready to sense the next row. This is known as a "precharge" operation, or "closing" the row. A precharge may be commanded explicitly, or it may be performed automatically at the conclusion of a read or write operation. Again, there is a minimum time, the row precharge delay, t_{RP}, which must elapse before that bank is fully idle and it may receive another activate command.

Although refreshing a row is an automatic side effect of activating it, there is a minimum time for this to happen, which requires a minimum row access time t_{RAS} delay between an Active command opening a row, and the corresponding precharge command closing it. This limit is usually dwarfed by desired read and write commands to the row, so its value has little effect on typical performance.

Command interactions

The no operation command is always permitted.

The load mode register command requires that all banks be idle, and a delay afterward for the changes to take effect.

The auto refresh command also requires that all banks be idle, and takes a refresh cycle time t_{RFC} to return the chip to the idle state. (This time is usually equal to $t_{RCD}+t_{RP}$.)

The only other command that is permitted on an idle bank is the active command. This takes, as mentioned above, t_{RCD} before the row is fully open and can accept read and write commands.

When a bank is open, there are four commands permitted: read, write, burst terminate, and precharge. Read and write commands begin bursts, which can be interrupted by following commands.

Interrupting a read burst

A read, burst terminate, or precharge command may be issued at any time after a read command, and will interrupt the read burst after the configured CAS latency. So if a read command is issued on cycle 0, another read command is issued on cycle 2, and the CAS latency is 3, then the first read command will begin bursting data out during cycles 3 and 4, then the results from the second read command will appear beginning with cycle 5.

If the command issued on cycle 2 were burst terminate, or a precharge of the active bank, then no output would be generated during cycle 5.

Although the interrupting read may be to any active bank, a precharge command will only interrupt the read burst if it is to the same bank or all banks; a precharge command to a different bank will not interrupt a read burst.

To interrupt a read burst by a write command is possible, but more difficult. It can be done, if the DQM signal is used to suppress output from the SDRAM so that the memory controller may drive data over the DQ lines to the SDRAM in time for the write operation. Because the effects of DQM on read data are delayed by 2 cycles, but the effects of DQM on write data are immediate, DQM must be raised (to mask the read data) beginning at least two cycles before write command, but must be lowered for the cycle of the write command (assuming you want the write command to have an effect).

Doing this in only two clock cycles requires careful coordination between the time the SDRAM takes to turn off its output on a clock edge and the time the data must be supplied as input to the SDRAM for the write on the following clock edge. If the clock frequency is too high to allow sufficient time, three cycles may be required.

If the read command includes auto-precharge, the precharge begins the same cycle as the interrupting command.

SDRAM burst ordering

A modern microprocessor with a cache will generally access memory in units of cache lines. To transfer a 64-byte cache line requires 8 consecutive accesses to a 64-bit DIMM, which can all be triggered by a single read or write command by configuring the SDRAM chips, using the mode register, to perform 8-word bursts.

A cache line fetch is typically triggered by a read from a particular address, and SDRAM allows the "critical word" of the cache line to be transferred first. ("Word" here refers to the width of the SDRAM chip or DIMM, which is 64 bits for a typical DIMM.) SDRAM chips support two possible conventions for the ordering of the remaining words in the cache line.

Bursts always access an aligned block of BL consecutive words beginning on a multiple of BL. So, for example, a 4-word burst access to any column address from 4 to 7 will return words 4 7. The ordering, however, depends on the requested address, and the configured burst type option: sequential or interleaved. Typically, a memory controller will require one or the other.

When the burst length is 1 or 2, the burst type does not matter. For a burst length of 1, the requested word is the only word accessed. For a burst length of 2, the requested word is accessed first, and the other word in the aligned block is accessed second. This is the following word if an even address was specified, and the previous word if an odd address was specified.

For the sequential burst mode, later words are accessed in increasing address order, wrapping back to the start of the block when the end is reached. So, for example, for a burst length of 4, and a requested column address of 5, the words would be accessed in the order 5-6-7-4. If the burst length were 8, the access order would be 5-6-7-0-1-2-3-4. This is done by adding a counter to the column address, and ignoring carries past the burst length.

The interleaved burst mode computes the address using an exclusive or operation between the counter and the address. Using the same starting address of 5, a 4-word burst would return words in the order 5-4-7-6. An 8-word burst would be 5-4-7-6-1-0-3-2. Although more confusing to humans, this can be easier to implement in hardware, and is preferred by Intel microprocessors.

If the requested column address is at the start of a block, both burst modes return data in the same sequential sequence 0-1-2-3-4-5-6-7. The difference only matters if fetching a cache line from memory in critical-word-first order.

SDRAM mode register

Single data rate SDRAM has a single 10-bit programmable mode register. Later double-data-rate SDRAM standards add additional mode registers, addressed using the bank address pins. For SDR SDRAM, the bank address pins and address lines A10 and above are ignored, but should be zero during a mode register write.

The bits are M9 through M0, presented on address lines A9 through A0 during a load mode register cycle.

1. M9: Write burst mode. If 0, writes use the read burst length and mode. If 1, all writes are non-burst (single location).
2. M8, M7: Operating mode. Reserved, and must be 00.
3. M6, M5, M4: CAS latency. Generally only 010 (CL2) and 011 (CL3) are legal. Specifies the number of cycles between a read command and data output from the chip. The chip has a fundamental limit on this value in nanoseconds; during initialization, the memory controller must use its knowledge of the clock frequency to translate that limit into cycles.
4. M3: Burst type. 0 - requests sequential burst ordering, while 1 requests interleaved burst ordering.
5. M2, M1, M0: Burst length. Values of 000, 001, 010 and 011 specify a burst size of 1, 2, 4 or 8 words, respectively. Each read (and write, if M9 is 0) will perform that many accesses, unless interrupted by a burst stop or other command. A value of 111 specifies a full-row burst. The burst will continue until interrupted. Full-row bursts are only permitted with the sequential burst type.

Later (double data rate) SDRAM standards use more mode register bits, and provide additional extended mode registers. The register number is encoded on the bank address pins during the load mode register cycle. For example, DDR2 SDRAM has a 13-bit mode register, a 13-bit EMR1, and uses 5 bits in EMR2.

Auto refresh

It is possible to refresh a RAM chip by opening and closing (activating and precharging) each row in each bank. However, to simplify the memory controller, SDRAM chips support an "auto refresh" command, which performs these operations to one row in each bank simultaneously. The SDRAM also maintains an internal counter, which iterates over all possible rows. The memory controller must simply issue a sufficient number of auto refresh commands (one per row, 4096 in the example we have been using) every refresh interval (t_{REF} = 64 ms is a common value). All banks must be idle (closed, precharged) when this command is issued.

Low power modes

As mentioned, the clock enable (CKE) input can be used to effectively stop the clock to an SDRAM. The CKE input is sampled each rising edge of the clock, and if it is low, the following rising edge of the clock is ignored for all purposes other than checking CKE. As long as CKE is low, it is permissible to change the clock rate, or even stop the clock entirely.

If CKE is lowered while the SDRAM is performing operations, it simply "freezes" in place until CKE is raised again.

If the SDRAM is idle (all banks precharged, no commands in progress) when CKE is lowered, the SDRAM automatically enters power-down mode, consuming minimal power until CKE is raised again. This must not last longer than the maximum refresh interval t_{REF}, or memory contents may be lost. It is legal to stop the clock entirely during this time for additional power savings.

Finally, if CKE is lowered at the same time as an auto-refresh command is sent to the SDRAM, the SDRAM enters self-refresh mode. This is like power down, but the SDRAM uses an on-chip timer to generate internal refresh cycles as necessary. The clock may be stopped during this time. While self-refresh mode consumes slightly more power than power-down mode, it allows the memory controller to be disabled entirely, which commonly more than makes up the difference.

SDRAM designed for battery-powered devices offers some additional power-saving options. One is temperature-dependent refresh; an on-chip temperature sensor reduces the refresh rate at lower temperatures, rather than always running it at the worst-case rate. Another is selective refresh, which limits self-refresh to a portion of the DRAM array. The fraction which is refreshed is configured using an extended mode register. The third, implemented in Mobile DDR (LPDDR) and LPDDR2 is "deep power down" mode, which invalidates the memory and requires a full reinitialization to exit from. This is activated by sending a "burst terminate" command while lowering CKE.

Generations of SDRAM

SDR SDRAM (Single Data Rate synchronous DRAM)

This type of SDRAM is slower than the DDR variants, because only one word of data is transmitted per clock cycle (single data rate). But this type is also faster than its predecessors EDO-RAM and FPM-RAM which took typically 2 or 3 clocks to transfer one word of data.

DDR SDRAM (*DDR1*)

While the access latency of DRAM is fundamentally limited by the DRAM array, DRAM has very high potential bandwidth because each internal read is actually a row of many thousands of bits. To make more of this bandwidth available to users, a double data rate interface was developed. This uses the same commands, accepted once per cycle, but reads or writes two words of data per clock cycle. The DDR interface accomplishes this by reading and writing data on both the rising and falling edges of the clock signal. In addition, some minor changes to the SDR interface timing were made in hindsight, and the supply voltage was reduced from 3.3 to 2.5 V. As a result, DDR SDRAM is not backwards compatible with SDR SDRAM.

DDR SDRAM (sometimes called *DDR1* for greater clarity) doubles the minimum read or write unit; every access refers to at least two consecutive words.

Typical DDR SDRAM clock rates are 133, 166 and 200 MHz (7.5, 6, and 5 ns/cycle), generally described as DDR-266, DDR-333 and DDR-400 (3.75, 3, and 2.5 ns per beat). Corresponding 184-pin DIMMs are known as PC-2100, PC-2700 and PC-3200. Performance up to DDR-550 (PC-4400) is available for a price.

DDR2 SDRAM

DDR2 SDRAM is very similar to DDR SDRAM, but doubles the minimum read or write unit again, to 4 consecutive words. The bus protocol was also simplified to allow higher performance operation. (In particular, the "burst terminate" command is deleted.) This allows the bus rate of the SDRAM to be doubled without increasing the clock rate of internal RAM operations; instead, internal operations are performed in units 4 times as wide as SDRAM. Also, an extra bank address pin (BA2) was added to allow 8 banks on large RAM chips.

Typical DDR2 SDRAM clock rates are 200, 266, 333 or 400 MHz (periods of 5, 3.75, 3 and 2.5 ns), generally described as DDR2-400, DDR2-533, DDR2-667 and DDR2-800 (periods of 2.5, 1.875, 1.5 and 1.25 ns). Corresponding 240-pin DIMMS are known as PC2-3200 through PC2-6400. DDR2 SDRAM is now available at a clock rate of 533 MHz generally described as DDR2-1066 and the corresponding DIMMs are known as PC2-8500 (also named PC2-8600 depending on the manufacturer). Performance up to DDR2-1250 (PC2-10000) is available for a price.

Note that because internal operations are at 1/2 the clock rate, DDR2-400 memory (internal clock rate 100 MHz) has somewhat higher latency than DDR-400 (internal clock rate 200 MHz).

DDR3 SDRAM

DDR3 continues the trend, doubling the minimum read or write unit to 8 consecutive words. This allows another doubling of bandwidth and external bus rate without having to change the clock rate of internal operations, just the width. To maintain 800–1600 M transfers/s (both edges of a 400–800 MHz clock), the internal RAM array has to perform 100–200 M fetches per second.

Again, with every doubling, the downside is the increased latency. As with all DDR SDRAM generations, commands are still restricted to one clock edge and command latencies are given in terms of clock cycles, which are half the speed of the usually quoted transfer rate (a CAS latency of 8 with DDR3-800 is 8/(400 MHz) = 20 ns, exactly the same latency of CAS2 on PC100 SDR SDRAM).

DDR3 memory chips are being made commercially,[2] and computer systems are available that use them as of the second half of 2007,[3] with expected significant usage in 2008.[4] Initial clock rates were 400 and 533 MHz, which are described as DDR3-800 and DDR3-1066 (PC3-6400 and PC3-8500 modules), but 667 and 800 MHz, described as DDR3-1333 and DDR3-1600 (PC3-10600 and PC3-12800 modules) are now common.[5] Performance up to DDR3-2200 (PC3 17600 modules) are available for a price.[6]

DDR4 SDRAM

DDR4 SDRAM will be the successor to DDR3 SDRAM. It was revealed at the Intel Developer Forum in San Francisco in 2008, and is due to be released to market during 2011. The timing has varied considerably during its development - it was originally expected to be released in 2012,[7] and later (during 2010) expected to be released in 2015,[8] before samples were announced in early 2011 and manufacturers began to announce that commercial production and release to market was anticipated in 2012. DDR4 is expected to reach mass market adoption around 2015, which is comparable with the approximately 5 years taken for DDR3 to achieve mass market transition over DDR2.

The new chips are expected to run at 1.2 V or less,[9] [10] versus the 1.5 V of DDR3 chips, and have in excess of 2 billion data transfers per second. They are expected to be introduced at frequency rates of 2133 MHz, estimated to rise to a potential 4266 MHz [11] and lowered voltage of 1.05 V [12] by 2013.

DDR4 will *not* double the internal prefetch width again, but will use the same 8*n* prefetch as DDR3.[13] Thus, it will be necessary to interleave reads from several banks to keep the data bus busy.

In February 2009, Samsung validated 40 nm DRAM chips, considered a "significant step" towards DDR4 development[14] since as of 2009, current DRAM chips were only beginning to migrate to a 50 nm process.[15] In

January 2011, Samsung announced the completion and release for testing of a 30 nm 2 GB DDR4 DRAM module. It has a maximum bandwidth of 2.13 Gbit/s at 1.2 V, uses pseudo open drain technology and draws 40% less power than an equivalent DDR3 module.[16] [17]

Feature map

Type	Feature changes
SDRAM	V_{cc} = 3.3 V Signal: LVTTL
DDR1	Access is ≥2 words Double clocked V_{cc} = 2.5 V 2.5 - 7.5 ns per cycle Signal: SSTL_2 (2.5V)[18]
DDR2	Access is ≥4 words "Burst terminate" removed 4 units used in parallel 1.25 - 5 ns per cycle Internal operations are at 1/2 the clock rate. Signal: SSTL_18 (1.8V)[18]
DDR3	Access is ≥8 words Signal: SSTL_15 (1.5V)[18] Much longer CAS latencies
DDR4	V_{cc} ≤ 1.2 V point-to-point (single module per channel)

Failed successors

In addition to DDR, there were several other proposed memory technologies to succeed SDR SDRAM.

Rambus DRAM (RDRAM)

RDRAM was a proprietary technology that competed against DDR. Its relatively high price and disappointing performance (resulting from high latencies and a narrow 16-bit data channel versus DDR's 64 bit channel) caused it to lose the race to succeed SDR DRAM.

Synchronous-Link DRAM (SLDRAM)

SLDRAM boasted higher performance and competed against RDRAM. It was developed during the late 1990s by the SLDRAM Consortium, which consisted of about 20 major computer industry manufacturers. It is an open standard and does not require licensing fees. The specifications called for a 64-bit bus running at a 200, 300 or 400 MHz clock frequency. This is achieved by all signals being on the same line and thereby avoiding the synchronization time of multiple lines. Like DDR SDRAM, SLDRAM uses an double-pumped bus, giving it an effective speed of 400,[19], 600,[20] or 800 MT/s.

SLDRAM used an 11-bit command bus (10 command bits CA9:0 plus one start-of-command FLAG line) to transmit 40-bit command packets on 4 consecutive edges of a differential command clock (CCLK/CCLK#). Unlike SDRAM, there were no per-chip select signals; each chip was assigned an ID when reset, and the command contained the ID of the chip that should process it. Data was transferred in 4- or 8-word bursts across an 18-bit (per chip) data bus, using one of two differential data clocks (DCLK0/DCLK0# and DCLK1/DCLK1#). Unlike standard SDRAM, the clock was generated by the data source (the SLDRAM chip in the case of a read operation) and transmitted in the same direction as the data, greatly reducing data skew. To avoid the need for a pause when the source of the DCLK changes, each command specified which DCLK pair it would use.[21]

The basic read/write command consisted of (beginning with CA9 of the first word):

SLDRAM Read, write or row op request packet

FLAG	CA9	CA8	CA7	CA6	CA5	CA4	CA3	CA2	CA1	CA0
1	ID8	Device ID							ID0	CMD5
0	Command code				CMD0	Bank			Row	
0	Row (continued)									0
0	0	0	0	Column						

- 9 bits of device ID
- 6 bits of command
- 3 bits of bank address
- 10 or 11 bits of row address
- 5 or 4 bits spare for row or column expansion
- 7 bits of column address

Individual devices had 8-bit IDs. The 9th bit of the ID sent in commands was used to address multiple devices. Any aligned power-of-2 sized group could be addressed. If the transmitted msbit was set, all least-significant bits up to and including the least-significant 0 bit of the transmitted address were ignored for "is this addressed to me?" purposes. (If the ID8 bit is actually considered less significant than ID0, the unicast address matching becomes a special case of this pattern.)

A read/write command had the msbit clear:

- CMD5=0
- CMD4=1 to open (activate) the specified row; CMD4=0 to use the currently open row
- CMD3=1 to transfer an 8-word burst; CMD3=0 for a 4-word burst
- CMD2=1 for a write, CMD2=0 for a read
- CMD1=1 to close the row after this access; CMD1=0 to leave it open
- CMD0 selects the DCLK pair to use (DCLK1 or DCLK0)

A notable omission from the specification was per-byte write enables; it was designed for systems with caches and ECC memory, which always write in multiples of a cache line.

Additional commands (with CMD5 set) opened and closed rows without a data transfer, performed refresh operations, read or wrote configuration registers, and performed other maintenance operations. Most of these commands supported an additional 4-bit sub-ID (sent as 5 bits, using the same multiple-destination encoding as the primary ID) which could be used to distinguish devices that were assigned the same primary ID because they were connected in parallel and always read/written at the same time.

There were a number of 8-bit control registers and 32-bit status registers to control various device timing parameters.

Virtual Channel Memory (VCM) SDRAM

VCM was a proprietary type of SDRAM that was designed by NEC, but released as an open standard with no licensing fees. VCM creates a state in which the various system processes can be assigned their own virtual channel, thus increasing the overall system efficiency by avoiding the need to have processes share buffer space. This is accomplished by creating different "blocks" of memory, allowing each individual memory block to interface separately with the memory controller and have its own buffer space. VCM has higher performance than SDRAM because it has significantly lower latencies. The technology was a potential competitor of RDRAM because VCM was not nearly as expensive as RDRAM was. A Virtual Channel Memory (VCM) module is mechanically and electrically compatible with standard SDRAM, but must be recognized by the memory controller. Few motherboards were ever produced with VCM support.

See also

- GDDR (graphics DDR) and its subtypes GDDR2, GDDR3, GDDR4, and GDDR5
- SDRAM latency
- List of device bandwidths
- Serial presence detect - EEPROM with timing data on SDRAM modules
- SDRAM Tutorial [22] - Flash website built by Tel-Aviv University students
- A concise but thorough review of SDRAM architecture/terminology and command timing dependencies in High-Performance DRAM System Design Constraints and Considerations [23], a master thesis from the University of Maryland.

References

[1] "SDRAM Part Catalog" (http://www.micron.com/products/dram/sdram/partlist). . 070928 micron.com
[2] "What is DDR memory?" (http://www.simmtester.com/page/news/showpubnews.asp?num=145). .
[3] Thomas Soderstrom (June 5, 2007). "Pipe Dreams: Six P35-DDR3 Motherboards Compared" (http://www.tomshardware.com/2007/06/05/pipe_dreams_six_p35-ddr3_motherboards_compared/). Tom's Hardware. .
[4] "AMD to Adopt DDR3 in Three Years" (http://news.softpedia.com/news/AMD-to-Adopt-DDR3-in-Three-Years-13486.shtml). .
[5] Wesly Fink (July 20, 2007). "Super Talent & TEAM: DDR3-1600 Is Here!" (http://www.anandtech.com/printarticle.aspx?i=3045). Anandtech. .
[6] Thomas Jørgen Jacobsen (28 July 2009). "A-Data launches DDR3-2200 with 2oz. copper PCB" (http://www.brightsideofnews.com/news/2009/7/28/a-data-launches-ddr3-2200-with-2oz-pcb.aspx). .
[7] DDR4 PDF page 23 (http://intel.wingateweb.com/US08/published/sessions/MASS006/SF08_MASS006_100s.pdf)
[8] http://www.semiaccurate.com/2010/08/16/ddr4-not-expected-until-2015/
[9] Looking forward to DDR4 (http://www.pcpro.co.uk/news/220257/idf-ddr3-wont-catch-up-with-ddr2-during-2009.html)
[10] DDR3 successor (http://www.heise-online.co.uk/news/IDF-DDR4-the-successor-to-DDR3-memory--/111367)
[11] "Next-Generation DDR4 Memory to Reach 4.266GHz - Report" (http://www.xbitlabs.com/news/memory/display/20100816124343_Next_Generation_DDR4_Memory_to_Reach_4_266GHz_Report.html). Xbitlabs.com. August 16, 2010. . Retrieved 2011-01-03.
[12] "IDF: DDR4 memory targeted for 2012" (http://www.hardware-infos.com/news.php?news=2332) (in German). hardware-infos.com. . Retrieved 2009-06-16. *English translation* (http://translate.google.com/translate?hl=en&sl=de&u=http://www.hardware-infos.com/news.php?news=2332&ei=bi44Sv_wBouZjAfVzYyjDQ&sa=X&oi=translate&resnum=1&ct=result&prev=/search?q=http://www.hardware-infos.com/news.php%3Fnews%3D2332&hl=en&safe=off&num=100)
[13] "JEDEC Announces Key Attributes of Upcoming DDR4 Standard" (http://www.jedec.org/news/pressreleases/jedec-announces-key-attributes-upcoming-ddr4-standard) (Press release). JEDEC. 2011-08-22. . Retrieved 2011-01-06.
[14] Gruener, Wolfgang (February 4, 2009). "Samsung hints to DDR4 with first validated 40 nm DRAM" (http://www.tgdaily.com/content/view/41316/139/). tgdaily.com. . Retrieved 2009-06-16.
[15] Jansen, Ng (January 20, 2009). "DDR3 Will be Cheaper, Faster in 2009" (http://www.dailytech.com/DDR3+Will+be+Cheaper+Faster+in+2009/article13977.htm). dailytech.com. . Retrieved 2009-06-17.
[16] "Samsung Develops Industry's First DDR4 DRAM, Using 30nm Class Technology" (http://www.samsung.com/us/business/semiconductor/newsView.do?news_id=1202). Samsung. 2011-01-04. . Retrieved 2011-03-13.
[17] http://www.techspot.com/news/41818-samsung-develops-ddr4-memory-up-to-40-more-efficient.html
[18] "EDA DesignLine, januari 12, 2007, The outlook for DRAMs in consumer electronics" (http://www.edadesignline.com/196900432?printableArticle=true). . 100622 edadesignline.com
[19] Dean Kent (1998-10-24), *RAM Guide: SLDRAM* (http://www.tomshardware.com/reviews/ram-guide,89-15.html), Tom's Hardware, , retrieved 2011-01-01
[20] Hyundai Electronics (1997-12-20), *HYSL8M18D600A 600 Mb/s/pin 8M x 18 SLDRAM* (http://icwic.cn/icwic/data/pdf/cd/cd011/12452.pdf) (data sheet), , retrieved 2011-12-27
[21] SLDRAM Inc. (1998-07-09), *SLD4M18DR400 400 Mb/s/pin 4M x 18 SLDRAM* (http://icwic.cn/icwic/data/pdf/cd/cd011/12407.pdf) (data sheet), pp. 32–33, , retrieved 2011-12-27
[22] http://taututorial.yolasite.com/
[23] http://drum.lib.umd.edu/bitstream/1903/11269/1/Gross_umd_0117N_11844.pdf

Infrared_Data_Association

The **Infrared Data Association** (**IrDA**) defines physical specifications communications protocol standards for the short-range exchange of data over infrared light, for uses such as personal area networks (PANs).

IrDA Logo

- IrDA is a very short-range example of free space optical communication.
- IrDA interfaces are used in medical instrumentation, test and measurement equipment, palmtop computers, mobile phones, and laptop computers (most laptops and phones also offer Bluetooth but it is now becoming more common for Bluetooth to simply replace IrDA in new versions of products).
- IrDA specifications include IrPHY, IrLAP, IrLMP, IrCOMM, Tiny TP, IrOBEX, IrLAN and IrSimple. IrDA has now produced another standard, IrFM, for Infrared financial messaging (i.e., for making payments) also known as "Point & Pay".

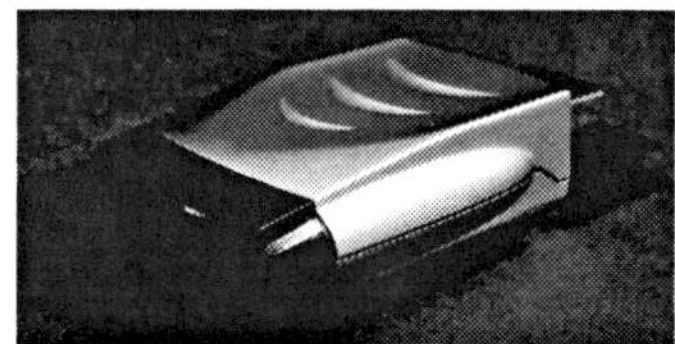

IrDA via USB

For the devices to communicate via IrDA, they must have a direct line of sight similar to a TV remote control.

IrDA in mobile phone Siemens CXT70

Specifications

IrPHY

The mandatory **IrPHY** (**Infrared Physical Layer Specification**) is the first (lowest) layer of the IrDA specifications. The most important specifications are:

- Range: standard: 1 m; low power to low power: 0.2 m; standard to low power: 0.3 m
- Angle: minimum cone ±15°
- Speed: 2.4 kbit/s to 1 Gbit/s
- Modulation: baseband, no carrier
- Infrared window
- Wavelength: 875 ± 30 nm[1]

IrDA transceivers communicate with infrared pulses (samples) in a cone that extends minimum 15 degrees half angle off center. The IrDA physical specifications require that a minimum irradiance be maintained so that a signal is visible up to a meter away. Similarly, the specifications require that a maximum irradiance not be exceeded so that a receiver is not overwhelmed with brightness when a device comes close. In practice, there are some devices on the market that do not reach one meter, while other devices may reach up to several meters. There are also devices that do not tolerate extreme closeness. The typical sweet spot for IrDA communications is from 5 to 60 cm (2.0 to 24 in) away from a transceiver, in the center of the cone. IrDA data communications operate in half-duplex mode because while transmitting, a device's receiver is blinded by the light of its own transmitter, and thus, full-duplex communication is not feasible. The two devices that communicate simulate full duplex communication by quickly turning the link around. The primary device controls the timing of the link, but both sides are bound to certain hard constraints and are encouraged to turn the link around as fast as possible.

Transmission rates fall into various categories: SIR, MIR, FIR, VFIR, UFIR, and Giga-IR. Serial Infrared (SIR) speeds cover those transmission speeds normally supported by an RS-232 port (9600 bit/s, 19.2 kbit/s, 38.4 kbit/s, 57.6 kbit/s, 115.2 kbit/s). Since the lowest common denominator for all devices is 9600 bit/s, all discovery and negotiation is performed at this baud rate. MIR (Medium Infrared) is not an official term, but is sometimes used to refer to speeds of 0.576 Mbit/s and 1.152 Mbit/s. Fast Infrared (FIR) is deemed an obsolete term by the IrDA physical specification, but is nonetheless in common usage to denote transmission at 4 Mbit/s. FIR is sometimes used to refer to all speeds above SIR. However, different encoding approaches are used by MIR and FIR, and different approaches frame MIR and FIR packets. For that reason, these unofficial terms have sprung up to differentiate these two approaches. The future holds faster transmission speeds (currently referred to as Very Fast Infrared, or VFIR) which supports a speed of 16 Mbit/s. There are (VFIR) infrared transceivers available such as the TFDU8108 operating from 9.6 kbit/s to 16 Mbit/s. The UFIR (Ultra Fast Infrared) protocol supports a speed of 96 Mbit/s. There, 8B10B coding is used. The Giga-IR protocol supports transmission speeds of 512 Mbit/s (64MB/s) and 1 Gbit/s (125 MB/s) and uses 2-ASK and 4-ASK modulation.

These are distinguished from other similar infrared communications systems that operate outside IrDA specifications. PDA communications formats such as HPSIR and ASKIR are such. This also includes CIR (Consumer IR), commonly used in remote controls, based on a raw protocol which uses sequences of pulse and space. It's possible to manage CIR via software like Lirc.

IrLAP

The mandatory **IrLAP (Infrared Link Access Protocol)** is the second layer of the IrDA specifications. It lies on top of the IrPHY layer and below the IrLMP layer. It represents the Data Link Layer of the OSI model. The most important specifications are:

- Access control
- Discovery of potential communication partners
- Establishing of a reliable bidirectional connection
- Distribution of the Primary/Secondary device roles
- Negotiation of QoS Parameters

On the IrLAP layer the communicating devices are divided into a Primary Device and one or more Secondary Devices. The Primary Device controls the Secondary Devices. Only if the Primary Device requests a Secondary Device to send is it allowed to do so.

IrLMP

The mandatory **IrLMP (Infrared Link Management Protocol)** is the third layer of the IrDA specifications. It can be broken down into two parts. First, the LM-MUX (Link Management Multiplexer) which lies on top of the IrLAP layer. Its most important achievements are:

- Provides multiple logical channels
- Allows change of Primary/Secondary devices

Second, the LM-IAS (Link Management Information Access Service), which provides a list, where service providers can register their services so other devices can access these services via querying the LM-IAS.

Tiny TP

The optional **Tiny TP** (**Tiny Transport Protocol**) lies on top of the IrLMP layer. It provides:

- Transportation of large messages by SAR (Segmentation and Reassembly)
- Flow control by giving credits to every logical channel

IrCOMM

The optional **IrCOMM** (**Infrared Communications Protocol**) lets the infrared device act like either a serial or parallel port. It lies on top of the IrLMP layer.

OBEX

The optional **OBEX** (**Object Exchange**) provides the exchange of arbitrary data objects (e.g., vCard, vCalendar or even applications) between infrared devices. It lies on top of the Tiny TP protocol, so Tiny TP is mandatory for OBEX to work.

IrLAN

The optional **IrLAN** (**Infrared Local Area Network**) provides the possibility to connect an infrared device to a local area network. There are three possible methods:

- Access Point
- Peer to Peer
- Hosted

As IrLAN lies on top of the Tiny TP protocol, the Tiny TP protocol must be implemented for IrLAN to work.

IrSimple

IrSimple achieves at least 4 to 10 times faster data transmission speeds by improving the efficiency of the infrared IrDA protocol. A 500 KB normal picture from a cell phone can be transferred within 1 second.[2]

IrSimpleShot

One of the primary targets of **IrSimpleShot (IrSS)** is to allow the millions of IrDA-enabled camera phones to wirelessly transfer pictures to printers, printer kiosks, flat panel TV's.

Popularity

IrDA was popular on PDA's, laptops and some desktops during the late 90s through the early 2000s. However, it has been displaced by other wireless technologies such as WiFi and Bluetooth, favored because they don't need a direct line of sight, and can therefore support hardware such as mice and keyboards. It is still used in some environments where interference makes radio-based wireless technologies unusable. IrDA popularity is making a comeback with its highly efficient IrSimple protocols by providing sub 1 second transfers of pictures between cell phones, printers, and display devices.[2] IrDA hardware is still less expensive and doesn't share the same security problems encountered with wireless technologies such as Bluetooth. In addition, some Pentax DSLRs (K-x, K-r) use IRsimple for image transfer and gaming[3] .

See also

- List of device bandwidths
- Consumer IR
- RZI

External links

- Infrared Data Association website [4]
- IrPro IrDA Protocol Stacks [5]
- ZMD IRDA [6]
- Linux Infrared HOWTO [7]
- Linux Infrared Remote Control [8]
- Linux status of infrared devices (IrDA, ConsumerIR, Remote Control) [9]
- Latest IRDA developments [10] including IrSimple, VFIR and UFIR (2005)
- IrDA project of Universidad Nacional de Colombia SIE board [11]

References

[1] (http://www.faculty.iu-bremen.de/birk/lectures/PC101-2003/17bluetooth/bluetooth/irda.html)
[2] IrDA IrSimple Specifications (Infrared Data Association - irda.org) (http://www.irda.org/displaycommon.cfm?an=1&subarticlenbr=48)
[3] (http://www.pentaximaging.com/about-us.aspx?p=press&pid=PENTAXANNOUNCESK-rDIGITALSLRANDNEW35MMLENS20100908174223) Pentax K-r
[4] http://www.irda.org/
[5] http://www.embednet.com
[6] http://www.zmd.biz/irda.php?group=irda&content=about
[7] http://www.tldp.org/HOWTO/Infrared-HOWTO/
[8] http://www.lirc.org
[9] http://tuxmobil.org/ir_misc.html
[10] http://www.deviceforge.com/news/NS7351692307.html
[11] http://wiki.linuxencaja.net/wiki/Academico/Embebidos/IrDA_SIE

- Charles D. Knutson with Jeffrey M. Brown, IrDA Principles and Protocols, 2004, ISBN 0-9753892-0-3

Series_40

Series 40 is a software platform and application user interface (UI) software on Nokia's broad range of mid-tier feature phones, as well as on the Vertu line of luxury phones. It is the world's most widely used mobile phone platform and found in hundreds of millions of devices.[1] Analyst estimates place the total number of devices shipped with S40 close to 1.5 billion, as of the end of 2010.[2]

Series 40-based Nokia 6300

Features

Applications

It provides communication applications such as telephone, internet telephony (VoIP), messaging, email client with POP3 and IMAP4 capabilities and Web browser; media applications such as camera, video recorder, music/video player and FM radio; and phonebook and other personal information management (PIM) applications such as calendar and tasks. Basic file management, like in Series 60, is provided in the Applications and Gallery folders and subfolders. Gallery is also the default location for files transferred over Bluetooth to be placed. User-installed applications on Series 40 are generally mobile Java applications. Flash Lite applications are also supported, but mostly used for screensavers.[3]

Web browser

The integrated web browser can access most web content through the service provider's XHTML/HTML gateway. The latest version of Series 40, called Series 40 6th Edition, introduced a new browser based on the WebKit open source components WebCore and JavaScriptCore. The new browser delivers support for HTML 4.01, CSS2, JavaScript 1.5, and Ajax. Also, like the higher-end Series 60, Series 40 can run the Opera Mini web browser to enhance the user's web browsing experience.

Synchronization

Support for SyncML synchronization with external services of the address book, calendar and notes is present. However with many S40 phones, these synchronization settings must be sent via an OTA text message.

Technical

Software platform

Series 40 is an embedded software platform that is open for software development via standard or de-facto content and application development technologies. It supports Java MIDlets, i.e. Java MIDP and CLDC technology, which provide location, communication, messaging, media, and graphics capabilities.[4] S40 also supports Flash Lite applications.[3]

Operating system

Series 40 is a simpler operating system than the higher end S60 (which is based on the multi-tasking Symbian OS). Because S40 devices do not support true multi-tasking and do not have a native code API for third parties, its user interface may appear to be more responsive and faster than other Nokia platforms on similar hardware.[5]

History

Series 40 was officially introduced in 2002 with release of Nokia 7210 [6]. It had 128x128 pixel color display (4096 colors) and other new features, such as MMS, J2ME, and MIDI ring tones. Over the years, the S40 UI has evolved from a low resolution UI to a high resolution color UI with an enhanced graphical look. The third generation of Series 40 that became available in 2005 introduced support for devices with resolutions as high as QVGA (240x320).[7] It is possible to customize the look-and-feel of the UI via comprehensive themes.[8] A list of all Series 40 devices can be found on the Nokia web site.[9]

See also

- Symbian
- Maemo
- MeeGo

References

[1] "Forum Nokia - Nokia Series 40 Platform" (http://www.forum.nokia.com/Devices/Series_40/). Nokia. . Retrieved 2010-10-27.

[2] "100 Million Club H2 2010" (http://www.100millionclub.com). VisionMobile. . Retrieved 2011-05-24.

[3] "Working with Nokia Series 40 Flash Lite content – Adobe Developer Center" (http://web.archive.org/web/20080518220200/http://www.adobe.com/devnet/devices/articles/nokia_series40_pt1_print.html). Adobe Systems. Archived from the original (http://www.adobe.com/devnet/devices/articles/nokia_series40_pt1_print.html) on 2008-05-18. . Retrieved 2008-09-26.

[4] "Developing Scalable Series 40 Applications, A Guide for Java Developers" (http://www.pearsoned.co.uk/BOOKSHOP/detail.asp?item=100000000075919). Addison-Wesley. . Retrieved 2008-09-26.

[5] "Comparing Series 40 against S60 (as of 2007) – All about Symbian" (http://www.allaboutsymbian.com/features/item/Series_40_vs_S60.php). . Retrieved 2008-09-26.

[6] http://press.nokia.com/2002/03/12/nokia-introduces-a-fashionable-tri-band-phone/

[7] "Series 40 UI Style Guide – Forum Nokia" (http://www.forum.nokia.com/info/sw.nokia.com/id/73e935fe-8b59-43b2-ab3e-1c5f763672db/Series_40_UI_Style_Guide.html). Nokia. . Retrieved 2008-09-26.

[8] "Carbide.ui Theme Edition (can be used to create S40 themes) – Forum Nokia" (http://www.forum.nokia.com/Library/Tools_and_downloads/Other/Carbide.ui/). Nokia. . Retrieved 2011-05-16.

[9] "Series 40 Platform SDKs" (http://www.forum.nokia.com/Library/Tools_and_downloads/Other/Series_40_platform_SDKs/). Nokia. . Retrieved 2010-09-19.

Article Sources and Contributors

Nokia_9500_Communicator *Source*: http://en.wikipedia.org/w/index.php?title=Nokia_9500_Communicator *Contributors*: -Majestic-, Adrianwn, Akash 2k07, Aletheides, Andros 1337, BentleyCoon, Bluemoose, Bobak, Bogdangiusca, Cacophony, Cyrius, Djr xi, E Wing, Edward, Gueneverey, Guy Harris, Harryson, Heman, Jack007, Julekmen, KC., Kevloral, Knewid, Mild Bill Hiccup, MrOllie, Mwanner, Newone, Niteowlneils, Od Mishehu, Open2universe, Orderud, Peter Grey, Petri Krohn, Pol098, Polluks, Sabre23t, Searchmaven, Smooth O, Squids and Chips, Template namespace initialisation script, Towel401, Treekids, Typ932, Uzume, Vespristiano, VikOlliver, Vitz-RS, Vlad, Wibbble, Yonkie, Zocky, Zpetro, 37 anonymous edits

Smartphone *Source*: http://en.wikipedia.org/w/index.php?title=Smartphone *Contributors*: 16@r, 3Coins, A Man In Black, A123a10, A8UDI, AKA MBG, ALSLUG, AVM, Aaditya 7, Acalamari, Acather96, Aclassifier, Acolin f, Adriatikus, Aeons, Ahoerstemeier, Aillema, Akuyume, Alansohn, Alex890, Alf Boggis, Alf.laylah.wa.laylah, AlistairMcMillan, Allen3, Amarendra.Avinash, AndrewSpec, Andries, Andros 1337, Angela, AniRaptor2001, Ansible, Anupam, Arab Dynasty, Arafael, Arny, Arp120, Arunsingh16, Atama, Atenyi, Atul.ecn, Avoided, BD2412, Bahaltener, Bakshi41c, Balazer, Baronnet, Beland, Bendes, Bhny, Bibijee, Biker Biker, BillHaywood, BlkStarr, Bomazi, Bovineone, Branddobbe, Brianreading, Broccoli, CLW, CRJO-CRJO, Cactusframe, Cadiomals, Caj27, Calcwiki, CalumCook234, CanadianPenguin, Canalteen, CatherineMunro, Cellcom, ChaChaFut, Changshui88, Charivari, Chris the speller, Ciphers, Citizensmith, Cleared as filed, Closedmouth, Clovis Sangrail, Cmf, Cmlewan, Cmp101, Codetiger, Comaleaf, CommonsDelinker, Commontimect, Cool Hand Luke, Coolaaron88, CoolingGibbon, Cornea503, Cst17, Currab, Cxtom, Cyruslei, DaisyField, Dale Arnett, Dan6hell66, Daniel.finnan, DaveChild, Dcljr, Dcxf, Ddavid2005, Deane@gooroos.com, Delphii, DerBorg, Deviceapps, Diannaa, Diego Moya, DivineAlpha, Dmarquard, Dreambringer, Drobatch, E.au, EVula, EWikist, Editor182, Editrrr, Elassint, Electronicguru1, Emma23 K, Epolk, Eraserhead1, Erc, Erianna, EugeneZelenko, Eugenia loli, Evice, Evilruletheworld96, Evolutiondb, F1MotoGPWRC, FCYTravis, Farolif, Father Goose, Fcassia, Feinoha, Feydey, Fhontoy, Flowanda, Fourdee, Furrybeagle, Fuzheado, GTBacchus, Galadh, Garant^ ^, GateKeeper, Geologyguy, Georgy90, Giftlite, Giggett, Gilliam, Ginsengbomb, Giraffedata, Glacialfox, Glennwells, Gmumru, Gogo Dodo, GoingBatty, Gold Hat, Gomm, Gookey, Gordon Ecker, Graemel., Grahamperrin, Grajales, Graywolf, Graywolfmoon1, Gregorysalt, Greswik, Groink, Gsarwa, Guy Harris, Gwernol, HDCase, Ha us 70, Haakon, Hadiceberg, Halloween.mac, Hamiltha, Hammersoft, Hardylane, Harmil, Harryzilber, Haseo9999, Hede2000, Henrik, HereToHelp, Hervegirod, Hobartimus, Hu12, Human.v2.0, Huntersquid, HuskyMoon, HybridBoy, Hydrogen Iodide, Hydrox, Hypnotist uk, I already forgot, I, Podius, Ianb, Ianboudreault, Ianereed, Icedwater, Ignis Fatuus, Illegal Operation, Illyria05, Im Buff, Immunmotbluescreen, Imroy, Imsilly123, In2thats12, Indefatigable, Interframe, InternetMeme, Intershark, Ipsign, Irky, Ishdarian, ItsZippy, J, J.delanoy, JCDenton2052, JCrue, JNW, Jagdterrier, JamesWeb, JamieS93, Jantangring, Jeffq, Jekyllhide, Jerome Charles Potts, Jerryobject, Jfdwolff, Jim.henderson, Jimmy Bergmark, JoeSmack, John of Reading, JonHarder, Jonabbey, Jondel, JonoP, Joseph Solis in Australia, Josh Parris, Jostake, Jsherwood0, Jsribeiro, Juan M. Gonzalez, Jusdafax, Just another editor, Justin Mauger, Jwojdylo, Jæs, Kajowi, Kariteh, KarmaGeddon, Katieh5584, Kbdank71, Kcomstock, Kellen`, KelleyCook, Kentynet, Kevin Dorner, Kingpin13, Kinst, Klokbaske, KnowBuddy, Koavf, Kozuch, Kraftlos, Kresp0, Kudpung, KungFuMonkey, Kuru, Kwaichi, LaVieEntiere, Lai888, Lambtron, Larkhill97, Laurasmith70308, Lazulilasher, LeilaniLad, Lenin1991, Lester, Leszek Jańczuk, Leujohn, Level plus, Lewispb, Lifes g00d 561, Lilac Soul, Limequat, Little Mountain 5, Little Professor, Logan, Loser997, Lotje, LrdChaos, Lukan4ica, Lun Esex, Lun4tic, MER-C, MMuzammils, Mac, Mac John Concord, Magnus.de, Malik Shabazz, Mange01, Maniacgeorge, Manop, Manway, Marcisjustice, Marco.difresco, Marcus Qwertyus, Mark Kim, Marksza, Marrowmonkey, Martarius, Masgatotkaca, Mathiastck, Maury Markowitz, Maxí, Mcduck, Mckote, Mdrejhon, Mdwh, Meehawl, Meelar, Memaster3, Mephistophelian, MetaManFromTomorrow, Miaow Miaow, Mikael Häggström, Mike Dillon, Mike Linksvayer, Mikehelms, Mild Bill Hiccup, Minghong, Miquonranger03, Mix Bouda-Lycaon, Mlg07e, Moberg, ModWilliam, Monaarora84, Mononomic, Morphh, Mortense, Mosmof, Mr. Strong Bad, MrOllie, Mvjs, My76Strat, MySchizoBuddy, MyronAub, Myscrnnm, Nakakapagpabagabag, Nathan94124, Navacell, Nazrich, Ne0Freedom, Nealmcb, Neo Ogilvy, Neoarchon, Neoguy999115, Nerdeff, Netrapt, Nick Number, Nickolai 420, Night Gyr, Nightscream, Nitro.ajb, Nixdorf, Nneonneo, Nopetro, Nurg, Obey, Ohnoitsjamie, Oknazevad, OlavN, Old port, Oli Filth, Omegatron, Oomay, OpenToppedBus, P.Marlow, PS., PTSE, Paranntachin, Patrick, Pats1, Patvac-chs, Paul 1953, Pbrown111, Pdahomepage, Petershank, Peyre, Phatalflaw, Phatom87, Phearson, Philip Trueman, Piano non troppo, Pingveno, Piotrek54321, Pluma, Pmarshal, Pmlineditor, Pol098, Pol430, Poor Yorick, Posix memalign, Prathameshsasane, Prillen, Pvanheus, Pyfan, Quebec99, Querencia, R'n'B, RA0808, Radiier, Radon210, Rafael.sp, Ramu50, RedWolf, Reliablesoft, Repetition, Res2216firestar, ReverseEngineered, Reyk, Rgreed, Rhobite, Riadlem, Rich Farmbrough, Richardguk, Richiekim, Riki, Rollins83, Ron2, Ronz, Ruchir257, Rursus, SF007, SPQRobin, Sadegh87, Sainath468, Samad120, Samwb123, Sandstein, Sapibobo, Sayid98, Schrödinger's Cake, ScottyBerg, Scrtcwlvl, Sdfisher, Sdrazfar, Sean13zz, Sebastian Mandrean, Sebindcruz, Secretsorry, Serg3d2, Sfacets, Sfan00 IMG, Sfsmartp, Shalroth, Shawnc, Sheehan, Signalhead, Siliconov, Singerdg1, SirJibby, Skatebiker, Skintigh, Skittle, Slitchfield, Smart1954, Smmgeek, Snigbrook, Soliloquial, SpaniardGR, Spellmaster, Starofwonder, Stephen B Streater, Stephen Turner, Stjson, SuperHamster, Surenkarapetyan, Suwatest, Swoof, TVS99, Taka76, Takamaxa, Taketa, Tangent1000, Taskinen, TastyPoutine, Tatterfly, Technopat, Tesi1700, Tfgbd, ThaWhistle, ThaddeusB, The Pink Oboe, The Thing That Should Not Be, The Wild Falcon, The-apathy, Theaveng, Thestick, Thiled, Thisma, Thumperward, Ticell, Tide rolls, Tikru8, TimSE, Tklaer, Tmuller2, Tobias Bergemann, Togopogo, Tokyogamer, Tomas.turek18, Tombomp, Tommy2010, Tommytentimes, Tony1, Toussaint, Tpbradbury, Trasz, Treekids, Tuju, UCLATre, UnrealG, Urod, Uzume, Vegaswikian, Verkinto, Verne Equinox, Versus22, Vkem, Vrenator, Waqas Hussain, Westie4321, WhisperToMe, Whkoh, WiZZLa, Wideangle, Wiki admi, Wikimhb, WikipedianYknOK, Wikipelli, Winged-stone, Wireless Buddy, Wknight94, Woohookitty, XGrape, Xajel, Xrobertcmx, Yean3d, Youxiarock, Yug, Yunshui, Zach.vega1, Zanter, ZimZalaBim, ZipoB, Zpetro, 1098 anonymous edits

Series_90_(software_platform) *Source*: http://en.wikipedia.org/w/index.php?title=Series_90_%28software_platform%29 *Contributors*: Claunia, Davewho2, FleetCommand, GTBacchus, GeneralAntilles, Interiot, KAMiKAZOW, KC., Kevloral, Korrawit, Leandrod, Letdorf, Lkt1126, Longhair, Lost.goblin, Lproven, MMuzammils, The Seventh Taylor, TigerK 69, Vegaswikian, 12 anonymous edits

OMAP *Source*: http://en.wikipedia.org/w/index.php?title=OMAP *Contributors*: Abdull, Ahunt, Ailuros21, Alexander.stohr, Alxx, Amitkarpe, Amreshk, Atomicthumbs, AvicAWB, Azulene, Babub, Beej, Bender235, Bgkwtnyqhzor, Billgordon1099, Bilrand, Blitterbug, Bogdan Doandes, Bounce1337, Caiaffa, Cecilyen, Choi yoon-su, Chris Chittleborough, ChrisRing, Codepro, CooperKoepp, Cyrilgermond, DSP-user, Dethme0w, Dhylands, Dingo aus, Dstrobl, Econtechie, Eeekster, ElComandanteChe, Elektrokapuściocha, Esebi95, Ethd, FC360, Failed audit, Fairseeder, Fleminra, Flibble, Flominator, Fox hyx, Gamester17, Gardar Rurak, Gbruin, GeneralAntilles, Ginamccall, Gjs238, GregoireGentil, Gu1dry, Huangcjz, Hydrargyrum, Imroy, Ingvarr, Ioliver, IrfanFaiz, JLD, JadonK, Jantangring, Jayabharathg, Jerry.waller, Jerryobject, Jevansen, Jfromcanada, Jhumbo, Josiahritchie, Kaulike, Kcl48, Kenny goo, Kevloral, Komal shah802003, Kozuch, Letdorf, MER-C, MX44, Mahjongg, Manjunath500, Mars9820, Mrzaius, Nandcon, Nealmcb, Ngunasekera, NiTenIchiRyu, Nixdorf, O2bal17, Olivecc, Onightly, Ossguy, Oyot22, Panze, Pgan002, Pj nielsen, Pngolla, Polyparadigm, R'n'B, Radiant!, Raymondzzd, Reisio, Rhishi.joshi, Rilak, Robert The Rebuilder, SF007, Sebastimau, Sigmundur, Sinkpad, Svenrichard, Syedakbar, Tehsusenoh, Tgok, Thumperward, Tlse31, Trevj, User931, Wernher, Y3Ah27, Yannick56, Yclinda666, 251 anonymous edits

Synchronous_dynamic_random-access_memory *Source*: http://en.wikipedia.org/w/index.php?title=Synchronous_dynamic_random-access_memory *Contributors*: Alereon, ArchStanton69, ArielGold, Ash73, AxelBoldt, B4hand, Biker Biker, Blound, Borgx, C 1, CAkira, CCFreak2K, Cacadril, Capsitan, Ceros, Chepry, CosineKitty, Curtains01, Dabomb87, DannyKitty, Darin-0, Ddrane, Deanpcmad, Debresser, Denniss, Dgessner, DmitryKo, Donhoraldo, Dozens, Dr Border, DragonHawk, E2eamon, ELCleanup, Edokter, Electron9, Emijrp, Engineerism, Eranhi, Eternal nirvana21, FT2, Frap, GreyCat, Haham hanuka, Ham Pastrami, Hdante, Hecline3rd, Heresiarch, Hlinasp, Hmrox, Intgr, JCDenton2052, Jamelan, Jamesjv, Jared Preston, Javalenok, Jeffq, Jezaraknid, Jhw38, Jimmi Hugh, JohnOwens, Jshayden, Jts888, Karam.Anthony.K, Kigali1, Krawi, Krazymike, Kveri, Kyleess123, Landofjacks, Lanny.ch, LordBlizzard, LorenzoB, ManiacK, Marek69, Matejhowell, Matt Britt, Mindmatrix, Monty845, Mr Stephen, Music Sorter, NewEnglandYankee, Nikpapag, Oscillon, Oxymoron83, Pbb, Peter17, Pitel, Rastapoulous, Rich Farmbrough, Rilak, Seaphoto, Slakr, Smb1001, SpartanPhalanx, Squantmuts, Sterremix, Stillwaterising, Syp, Tcnuk, Tehom2000, Terryeo, Tgr, The toiletbrush, Thewikigod12345, Thunderbird2, Tom strom, Tony Fox, Trolleytimes, Wadsworth, Why Not A Duck, WillMak050389, WillWare, WorkingBeaver, Wtshymanski, ZZninepluralZalpha, Zhangxc, 253 anonymous edits

Infrared_Data_Association *Source*: http://en.wikipedia.org/w/index.php?title=Infrared_Data_Association *Contributors*: Abune, Alansohn, Andries, Angerdan, Angerstein, Banjodog, Bobblewik, Bobgastineau, Bowlhover, BrianWilloughby, Brianski, BryanG, Bumper12, CALR, CesarB, ClementSeveillac, DMacks, DaGizza, Dataphile, David Gerard, Ddcc, DerHexer, Dicklyon, Dondegroovily, Dr. Sunglasses, Eeekster, Engineerism, Eptin, Fishnet37222, Flewovercuckoo, Frank, Frappucino, Frehley, Graeme Bartlett, GrapeSmuckers, GreenUpGreenOut, Griphook, HarisM, Icairns, Imroy, Ing tes, Iuhkjhk87y678, Jaraalbe, Joel Saks, Joerg Reiher, Kenny sh, Kim SJ, Kozuch, Kurt Jansson, Kyng, Magus732, Manscher, Masamage, Materialscientist, Mayank gupta, Michael Hardy, Microzoft, Moanzhu, Mormegil, Oliverdl, Omicronpersei8, Poppafuze, Psychobunny2412, Qllach, Ramlalmani, Rror, Sanbec, Sega381, Splintercellguy, Suruena, Synthetik, Tero, The Anome, The Wordsmith, Thesundog, Tothwolf, Whpq, Wingman4l7, Тиверополник, 131 anonymous edits

Series_40 *Source*: http://en.wikipedia.org/w/index.php?title=Series_40 *Contributors*: Agreimann, Alex Perry, Aniac, Ashrutisingh, Azuya, Bjelleklang, Blakegripling ph, Calltech, Canaima, Centrx, Chapzboy, Charivari, Cnj, Daverocks, Diwas, Djlarz, Djr xi, Dktz, Dublinclontarf, Elspif, Ettrig, Evilboy, FleetCommand, Flod logic, Florin92, Gaius Cornelius, Hjorten, Imroy, JLaTondre, Jarekadam, Jasonon, Jesjimher, LordBen5000, MMuzammils, Mdwh, Milan292208, Monkeyhousetim, Msulik, Munamankeli, Park3r, Petri Krohn, Polisher of Cobwebs, Rich Farmbrough, Rlobkovsky, Roleplayer, Rudiedude, Rwwww, SimonMenashy, The Seventh Taylor, TheParanoidOne, Tirkfl, UKER, Vegaswikian, We hope, Xkury, Zakawer, ZeroOne, Zhernovoi, 97 anonymous edits

Image Sources, Licenses and Contributors

Image:Julie Carroll Moblogging.jpg *Source*: http://en.wikipedia.org/w/index.php?title=File:Julie_Carroll_Moblogging.jpg *License*: unknown *Contributors*: FlickrLickr, FlickreviewR, Jpk, MB-one

File:Group of smartphones.jpg *Source*: http://en.wikipedia.org/w/index.php?title=File:Group_of_smartphones.jpg *License*: unknown *Contributors*: gillyberlin

File:IBM SImon in charging station.png *Source*: http://en.wikipedia.org/w/index.php?title=File:IBM_SImon_in_charging_station.png *License*: unknown *Contributors*: User:Bcos47

File:Nokia 9210.jpg *Source*: http://en.wikipedia.org/w/index.php?title=File:Nokia_9210.jpg *License*: unknown *Contributors*: J-P Kärnä

File:Htc Touch Pro2 Georgy.JPG *Source*: http://en.wikipedia.org/w/index.php?title=File:Htc_Touch_Pro2_Georgy.JPG *License*: unknown *Contributors*: User:Georgy90

File:Original iPhone docked.jpg *Source*: http://en.wikipedia.org/w/index.php?title=File:Original_iPhone_docked.jpg *License*: unknown *Contributors*: Andrew from London, UK

File:Galaxy Nexus smartphone.jpg *Source*: http://en.wikipedia.org/w/index.php?title=File:Galaxy_Nexus_smartphone.jpg *License*: unknown *Contributors*: Faramarz, MB-one, SF007, 1 anonymous edits

File:Global Mobile Applications Store Revenue.svg *Source*: http://en.wikipedia.org/w/index.php?title=File:Global_Mobile_Applications_Store_Revenue.svg *License*: unknown *Contributors*: User:MySchizoBuddy

File:Smartphone share current.png *Source*: http://en.wikipedia.org/w/index.php?title=File:Smartphone_share_current.png *License*: unknown *Contributors*: -- Eraserhead1 <talk> 12:49, 3 March 2010 (UTC) Graph created by myself. Original uploader was Eraserhead1 at en.wikipedia

File:BeagleBoard described.jpg *Source*: http://en.wikipedia.org/w/index.php?title=File:BeagleBoard_described.jpg *License*: unknown *Contributors*: beagleboard.org

File:PandaBoard described.png *Source*: http://en.wikipedia.org/w/index.php?title=File:PandaBoard_described.png *License*: unknown *Contributors*: pandaboard.org

File:OMAP-850.jpg *Source*: http://en.wikipedia.org/w/index.php?title=File:OMAP-850.jpg *License*: unknown *Contributors*: Flibble

Image:SDR SDRAM-1.jpg *Source*: http://en.wikipedia.org/w/index.php?title=File:SDR_SDRAM-1.jpg *License*: unknown *Contributors*: Royan

Image:Micron 48LC32M8A2-AB.jpg *Source*: http://en.wikipedia.org/w/index.php?title=File:Micron_48LC32M8A2-AB.jpg *License*: unknown *Contributors*: Chepry, Jpk, MMuzammils, Solomon203

Image:Irdalogo.png *Source*: http://en.wikipedia.org/w/index.php?title=File:Irdalogo.png *License*: unknown *Contributors*: Bumper12

Image:IrDA USB.jpg *Source*: http://en.wikipedia.org/w/index.php?title=File:IrDA_USB.jpg *License*: unknown *Contributors*: Bartekbas

Image:IrDA w tel.jpg *Source*: http://en.wikipedia.org/w/index.php?title=File:IrDA_w_tel.jpg *License*: unknown *Contributors*: Bartekbas

Image:Nokia 6300.jpg *Source*: http://en.wikipedia.org/w/index.php?title=File:Nokia_6300.jpg *License*: unknown *Contributors*: Original uploader was Racklever at en.wikipedia

CPSIA information can be obtained at www.ICGtesting.com
Printed in the USA
LVOW041551220312
274341LV00004B/13/P